Nanophotonics

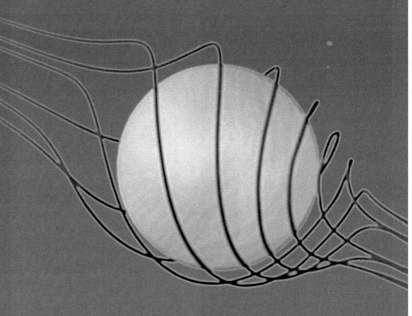

Nanophotonics
Manipulating Light with Plasmons

edited by
Hongxing Xu

PAN STANFORD PUBLISHING

Published by

Pan Stanford Publishing Pte. Ltd.
Penthouse Level, Suntec Tower 3
8 Temasek Boulevard
Singapore 038988

Email: editorial@panstanford.com
Web: www.panstanford.com

British Library Cataloguing-in-Publication Data
A catalogue record for this book is available from the British Library.

Nanophotonics: Manipulating Light with Plasmons
Copyright © 2018 by Pan Stanford Publishing Pte. Ltd.
All rights reserved. This book, or parts thereof, may not be reproduced in any form or by any means, electronic or mechanical, including photocopying, recording or any information storage and retrieval system now known or to be invented, without written permission from the publisher.

For photocopying of material in this volume, please pay a copying fee through the Copyright Clearance Center, Inc., 222 Rosewood Drive, Danvers, MA 01923, USA. In this case permission to photocopy is not required from the publisher.

ISBN 978-981-4774-14-7 (Hardcover)
ISBN 978-1-315-19661-9 (eBook)

Printed in the USA

Contents

Preface ix

1. Fundamentals of Plasmonics 1

Lianming Tong, Hong Wei, and Hongxing Xu

1.1	Introduction		1
1.2	Dielectric Function of Metal		2
1.3	Localized Surface Plasmons		5
	1.3.1	Quasi-Static Approximation	5
	1.3.2	Extinction, Scattering, and Absorption	7
	1.3.3	Near-Field Distribution	8
	1.3.4	Decay of Localized Surface Plasmons	9
1.4	Propagating Surface Plasmons		10
	1.4.1	Propagating Surface Plasmons at the Planar Metal–Dielectric Interface	10
	1.4.2	Length Scales of Propagating Surface Plasmons	12
1.5	Research Topics		14

2. Light Scattering by Small Metallic Particles: Mie Theory and More 21

Shunping Zhang and Hongxing Xu

2.1	Light Scattering by a Single Spherical Particle and Mie Theory		23
	2.1.1	Vector Solutions in the Spherical Coordinate: VSH	23
	2.1.2	Expansion of a Plane Wave by VSH	25
	2.1.3	Scattering by a Single Sphere	27
	2.1.4	Optical Far-Field Cross Sections	29
2.2	Generalized Mie Theory		31
	2.2.1	Scattering by a Multilayered Sphere	32
	2.2.2	Addition Theorem for VSH	35
	2.2.3	Order-of-Scattering Method for Two Spheres	37
	2.2.4	Order-of-Scattering Method for an Arbitrary Number of Spheres	40

2.3	Light Scattering by Arbitrarily Shaped Particles and Numerical Simulations	46
	2.3.1 The Green Dyadic Method	47
	2.3.2 Numerical Techniques	49
2.4	Summary	51

3. Electromagnetic Field Enhancement in Surface-Enhanced Raman Scattering   55

Ke Zhao, Hong Wei, and Hongxing Xu

3.1	Introduction	55
3.2	Numerical Approaches to EM Enhancement	57
3.3	The Nanogap Effect	60
3.4	Various Types of Nanogaps	68
3.5	Multiple-Particle Nanoantennas for Controlling Polarization of SERS Emission	76
3.6	Electronic Coupling in Nanogaps	77
3.7	Probing EM Enhancement via SERS	79
3.8	Summary	81

4. Plasmonic Antennas   85

Zhipeng Li, Longkun Yang, Hancong Wang, and Hongxing Xu

4.1	Introduction	85
4.2	Single Plasmonic Antennas	86
4.3	Coupled Optical Antennas	93
	4.3.1 Control of Local Intensity	93
	4.3.2 Control of Emission Direction	96
	4.3.3 Control of Far-Field Polarization	98
4.4	Summary	104

5. Plasmon-Assisted Optical Forces   109

Lianming Tong and Hongxing Xu

5.1	Introduction	109
5.2	Theoretical Calculations on Optical Forces in Near-Field-Coupled Nanoparticles	111
5.3	Experimental Demonstrations of Plasmon-Assisted Optical Forces	117

		5.3.1	Optical Forces on Metal Nanoparticles Trapped by a Focused Laser Beam	117
			5.3.1.1 Elongated nanoparticles	117
			5.3.1.2 Interaction between two metal nanoparticles in an optical trap	121
			5.3.1.3 Applications in SERS sensing	124
		5.3.2	Optical Forces in Lithographically Fabricated Plasmonic Nanostructures	127
			5.3.2.1 Gold nanopads and nanoholes	127
			5.3.2.2 Dimers of nanodisks	128
		5.3.3	Optical Forces in Propagating Surface Plasmon Systems: Gold Thin Films and Nanostripes	130
	5.4	Summary and Perspective		132

6. Plasmonic Nanowire Waveguides and Circuits — 137

Hong Wei and Hongxing Xu

	6.1	Introduction		137
	6.2	Excitation and Detection of Propagating SPPs		139
		6.2.1	SPP Excitation	139
		6.2.2	SPP Detection	141
	6.3	Fundamental Properties of SPPs in Metal Nanowires		145
		6.3.1	SPP Modes in Metal Nanowires	145
		6.3.2	SPP Propagation in Metal Nanowires	149
		6.3.3	Group Velocity	158
		6.3.4	Propagation Length and Loss	159
		6.3.5	Emission Direction and Polarization	164
		6.3.6	Spin–Orbit Interaction of Light in Plasmonic Nanowires	171
		6.3.7	Nanowire–Emitter Coupling	173
	6.4	Plasmonic Devices and Circuits		177
		6.4.1	SPP Router, Splitter, Demultiplexer, Switch, and Spin Sorter	177
		6.4.2	SPP Modulation, Logic Gates, and Computing	185
		6.4.3	Hybrid Plasmonic-Photonic Nanowire Devices	192
	6.5	Summary		194

7. Gain-Assisted Surface Plasmon Resonances and Propagation   **201**

Ning Liu and Hongxing Xu

7.1	Introduction	201
7.2	Amplification of Long-Range Surface Plasmon Polaritons	203
7.3	Stimulated Emission from Localized Surface Plasmon Resonance with a Gain Material	206
7.4	Gain-Assisted Hybrid Surface Plasmon Propagation: Lasing and Amplification	212
	7.4.1 Hybrid Surface Plasmon Lasers	212
	7.4.2 Amplification of Hybrid Surface Plasmon Polaritons	218
7.5	Summary and Future Perspective	221

Index   225

Preface

Manipulation of light at the nanometer scale is highly pursued for both fundamental sciences and wide applications. The related studies are within the scope of nanophotonics. The diffraction limit of light sets the limit for the smallest size of photonic devices to the scale of light wavelength. Fortunately, surface plasmons (SPs), collective oscillations of electrons at the surface of metal (mostly gold and silver) nanostructures, make it possible to squeeze light into nanoscale volumes and realize light manipulation beyond the diffraction limit.

SPs were discovered and formally named more than a half century ago, although the phenomena related to SPs appeared much earlier. They attracted renewed interest, and there was explosive growth in the 21st century due to the developments in nanostructure fabrications, characterizations, and simulations. The studies on SPs have formed a booming research field called plasmonics, which is concerned with the phenomena, mechanisms, devices, and applications based on SP resonances mostly in metallic and composite nanostructure systems.

Historically, two plasmon-related research topics predate the prosperity of plasmonics, surface-enhanced Raman scattering (SERS) and surface plasmon resonance (SPR) sensing. SERS, discovered in the 1970s, is caused by SP-induced electromagnetic field enhancement. Especially, the hugely enhanced electric field in the nanogap of a nanoparticle dimer reported in 1999 reveals clearly the mechanism of single-molecule SERS and demonstrates the most charming character of SPs in coupled metal nanostructures. The field enhancement effect in nanogaps has been widely used to enhance various light–matter interactions. The SPR sensing developed in the 1980s utilizes the SP resonances of metal films deposited on glass substrates. Their high sensitivity to the environmental change and easy fabrication procedure bring about the commercial application of SPR sensor chips.

Plasmon-based nanophotonics has potential applications in many fields, such as information technology, biological/chemical sensing, medical diagnosis and therapy, renewable energy, and super-resolution imaging. Field confinement beyond the diffraction limit enables the scaling down of photonic devices and facilitates the integration of nanophotonic devices with nanoelectronic devices. Due to its various potential applications, plasmonics attracts researchers in different fields, including physics, materials, biology, chemistry, and medicine, which makes plasmonics a truly interdisciplinary field.

The continuingly emerging research topics bring vitality to this field. For example, quantum plasmonics is developing quickly. On the one hand, quantum effects occur when the size of the nanogap is decreased to a few angstroms. On the other hand, quantum optics phenomena, such as quantum entanglement and quantum interference, are realized in plasmonic systems. Graphene plasmonics extends the plasmonic materials from metal to novel 2D materials. Plasmons have been involved in the studies of chemical reactions, leading to the creation of a new field, plasmochemistry. Plasmonic circular dichroism and various chiral and vortical effects form the field of chiral plasmonics. Photonic spin–orbit interactions and resulting novel optical phenomena in plasmonic nanostructures have been investigated recently.

This book gives a general introduction to the fundamental aspects of plasmonics. Chapter 1 presents a brief introduction to the fundamentals of localized SPs and propagating SPs and outlines the most active research topics. In Chapter 2, the computational approaches to light scattering due to SP resonances in metal nanoparticles are introduced. Chapters 3–5 cover three of the typical subfields resulting from SP resonances in metal nanoparticles: SERS, plasmonic nanoantennas, and plasmon-assisted optical forces. Chapter 6 focuses on propagating SPs in nanowire waveguides for controllable subwavelength light guiding and potential circuitry applications. Chapter 7 introduces the optical properties in plasmonic systems with gain materials.

Plasmonics is still developing fast, and new research topics keep emerging. This book is not intended to cover all aspects of plasmonics but gives some introduction to the fundamental and representative properties and research in this field, which can help graduate students and researchers get a quick study of this field and get the necessary information to step toward new research.

Hongxing Xu
Wuhan University
September 2016

Chapter 1

Fundamentals of Plasmonics

Lianming Tong,[a] Hong Wei,[b] and Hongxing Xu[c]

[a]*Center for Nanochemistry, College of Chemistry and Molecular Engineering, Peking University, Beijing 100871, China*
[b]*Institute of Physics, Chinese Academy of Sciences, Beijing 100190, China*
[c]*School of Physics and Technology, and Institute for Advanced Studies, Wuhan University, Wuhan 430072, China*
tonglm@pku.edu.cn; hxxu@whu.edu.cn

1.1 Introduction

Surface plasmons (SPs) are collective oscillations of conduction electrons in a metal surface, which can be excited by an electromagnetic (EM) wave [1, 2]. The splendid optical characteristics of SPs include, but are not limited to, preferential absorption and scattering of light of particular wavelengths, EM field confinement beyond the diffraction limit, local EM field enhancement, and long-range propagation and remote emission of EM waves, rendering a variety of important research topics in the field of plasmonics, such as surface-enhanced spectroscopy, surface plasmon resonance (SPR) sensing, plasmonic nanoantennas, enhanced photovoltaics, and plasmonic circuits, and enabling interesting applications

---

Nanophotonics: Manipulating Light with Plasmons
Edited by Hongxing Xu
Copyright © 2018 Pan Stanford Publishing Pte. Ltd.
ISBN 978-981-4774-14-7 (Hardcover), 978-1-315-19661-9 (eBook)
www.panstanford.com

in multidisciplinary fields involving physics, chemistry, biology, renewable energy, and information technology [3–9].

The eigenenergies of SPs in noble metals, typically gold and silver, are in the visible optical frequencies, making them popular materials in the research of SPs. The fascinating Lycurgus Cup in the ancient Roman age is probably the earliest demonstration of SP effects. The glass of the cup contains nanosized gold and silver particles and glows either red or green, depending on how light hits it. To understand the optical properties of metal nanoparticles, in the year 1857, Faraday successfully prepared pure colloidal gold and related its color to the small size of gold particles [10]. Later, in 1908, Mie developed a theory of solving Maxwell's equations to calculate the scattering and absorption of spherical nanoparticles [11].

Although it was developed more than 100 years ago, Mie theory has been one of the most important theoretical models to describe the optical properties of metal nanoparticles until now. Certainly, a series of electrodynamic simulation methods have also been developed, for example, the finite element method (FEM), finite-difference time-domain (FDTD), discrete dipole approximation (DDA), and the Green function method. The whole tool kits of modeling can now simulate almost all metal nanostructures—spherical/aspherical nanoparticles, nanorods, nanowires, nanoparticle clusters, nanoparticles in homogeneous/inhomogeneous media, etc. From the fundamental point of view, the optical response of a metal nanostructure is determined by several crucial physical parameters: the frequency-dependent complex permittivity of metal, the effective dielectric constant of surrounding media, and the geometric dimensions of the nanostructure. In this chapter, we briefly introduce some fundamentals of plasmonics.

1.2 Dielectric Function of Metal

For a better understanding of SPs, it is instructive to take a look at the optical properties of metals in a simplified picture. The free electrons in the conduction band collectively oscillate if illuminated by an external EM field. The motion of electrons can be simply regarded as a damped Lorentz oscillator. The motion equation of the free electrons with mass m and unit charge e in an external electric field $\mathbf{E}(t) = \mathbf{E}_0 \exp(-i\omega t)$ is as below:

$$m\frac{d^2\mathbf{x}}{dt^2} + m\gamma\frac{d\mathbf{x}}{dt} = -e\mathbf{E}_0\exp(-i\omega t), \tag{1.1}$$

where $\mathbf{x}$ is the position of an electron and γ is the damping coefficient. By solving the motion equation, we obtain the displacement of the electron oscillation:

$$\mathbf{x}(t) = \frac{e\mathbf{E}(t)}{m(\omega^2 + i\gamma\omega)}. \tag{1.2}$$

The macroscopic polarization is then written as

$$\mathbf{P} = N\cdot(-e\mathbf{x}) = \frac{-Ne^2\mathbf{E}(t)}{m(\omega^2 + i\gamma\omega)}, \tag{1.3}$$

where N is the number density of the free electrons. As the polarization also takes the form $\mathbf{P} = \varepsilon_0[\varepsilon(\omega) - 1]\mathbf{E}(t)$, upon substitution into Eq. 1.3 we can derive the complex dielectric function of metal:

$$\varepsilon(\omega) = \varepsilon' + i\varepsilon'' = 1 - \frac{\omega_p^2}{\omega(\omega + i\gamma)}, \tag{1.4}$$

$$\varepsilon' = 1 - \frac{\omega_p^2}{\omega^2 + \gamma^2}, \tag{1.5}$$

and

$$\varepsilon'' = \frac{\omega_p^2 \gamma}{\omega(\omega^2 + \gamma^2)}, \tag{1.6}$$

where $\omega_p = \sqrt{Ne^2/\varepsilon_0 m}$ is the bulk plasmon frequency, in which ε_0 is the vacuum permittivity, and ε' and ε'' are the real and imaginary parts, respectively, of $\varepsilon(\omega)$. This is the so-called Drude model. Although very simple, this model indeed successfully explains many of the optical properties of metals in the visible region.

For a quick glance, we consider the limit of low damping, where $\gamma \ll \omega$. Equations 1.5 and 1.6 are simplified to

$$\varepsilon' = 1 - \frac{\omega_p^2}{\omega^2}, \tag{1.7}$$

and

$$\varepsilon'' = \frac{\omega_p^2 \gamma}{\omega^3}. \tag{1.8}$$

The bulk plasmon frequencies for metals are typically in the ultraviolet, and for gold and silver ~8–10 eV. Now we see that in the visible light region where $\omega < \omega_p$, ε' is negative and ε'' also plays a role, indicating strong interaction between metal and the incident field. At frequencies $\omega > \omega_p$, ε' is positive and ε'' can be neglected, meaning that the metal acts as a dielectric and is transparent to the incident field. The reflection and refraction of light at the metal–air interface then obey Fresnel's law.

For noble metals, the d-band electrons that are close to the Fermi surface lead to a polarized background that contributes to the dielectric function. This is described in the Drude expression by the inclusion of ε_∞ in Eq. 1.4 and is written as

$$\varepsilon(\omega) = \varepsilon_\infty - \frac{\omega_p^2}{\omega(\omega + i\gamma)}. \tag{1.9}$$

Upon assumption that only the conduction electrons contribute, $\varepsilon_\infty = 1$, and Eq. 1.4 is recovered.

The Drude model does not consider interband transitions that contribute to damping and decrease the lifetime of plasmons. Figure 1.1 shows the real and imaginary parts of the dielectric function of gold calculated using the Drude model and experimentally measured by Johnson and Christy (JC) [12]. It is seen that the real part of the dielectric function agrees well between the Drude model and JC data, but the imaginary part differs significantly at high photon energies due to interband transitions.

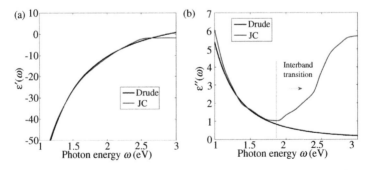

Figure 1.1 Real part (a) and imaginary part (b) of the dielectric function of gold plotted using the Drude model and JC data. Fitting parameters used in the Drude model are ω_p = 9.0 eV, ε_∞ = 9.8 eV, and γ = 67 meV. The imaginary part of the JC data is significantly larger than that in the Drude model due to interband transitions.

1.3 Localized Surface Plasmons

1.3.1 Quasi-Static Approximation

For nanoparticles, the electron oscillation is confined in three dimensions, resulting in the so-called localized surface plasmons. Considering a spherical metal nanoparticle of radius a in a static electric field $\mathbf{E}_0$ and ignoring the retardation effect, we can solve for the induced electric potential inside and outside the metal particle, as given below:

$$\Phi_{in} = \frac{-3\varepsilon_2}{\varepsilon + 2\varepsilon_2} E_0 r \cos\theta, \qquad (1.10)$$

and

$$\Phi_{out} = -E_0 r \cos\theta + a^3 E_0 \frac{\varepsilon - \varepsilon_2}{\varepsilon + 2\varepsilon_2} \frac{\cos\theta}{r^2}, \qquad (1.11)$$

where ε_2 is the dielectric constant of the surrounding medium and r and θ are position and angle parameters, respectively, of the point of interest. The first term in Eq. 1.11 is apparently from the incident electric field. By introducing an ideal dipole with dipole moment $\mathbf{p}$ at the center of the nanosphere, the second term can be rewritten as

$$a^3 E_0 \frac{\varepsilon - \varepsilon_2}{\varepsilon + 2\varepsilon_2} \frac{\cos\theta}{r^2} = \frac{\mathbf{p} \cdot \mathbf{r}}{4\pi\varepsilon_0 \varepsilon_2 r^3}, \qquad (1.12)$$

and

$$\mathbf{p} = 4\pi\varepsilon_0 \varepsilon_2 a^3 \frac{\varepsilon - \varepsilon_2}{\varepsilon + 2\varepsilon_2} \mathbf{E}_0. \qquad (1.13)$$

Then we see that the potential outside the particle is the contribution of the superposition of the incident field $\mathbf{E}_0$ and the field generated by a dipole with dipole moment $\mathbf{p}$, which can be defined with polarizability α as $\mathbf{p} = \varepsilon_0 \varepsilon_2 \alpha \mathbf{E}_0$. The polarizability α of this dipole can be easily found as (sometimes also called the Clausius–Mossotti polarizability)

$$\alpha = 4\pi a^3 \frac{\varepsilon - \varepsilon_2}{\varepsilon + 2\varepsilon_2}. \qquad (1.14)$$

The polarizability of a metal nanoparticle in the dipole approximation is a function of the dielectric constant of the metal and the surrounding medium and is also a complex quantity. The above model is derived on the basis of a static electric field. When applied to an incident EM wave that has a time-varying electric field, it is also referred to as quasi-static approximation.

From Eq. 1.14, it is straightforward that the dipolar resonance is achieved by minimizing the denominator $|\varepsilon + 2\varepsilon_2|$. Considering the complex dielectric constant of metal $\varepsilon(\omega) = \varepsilon' + i\varepsilon''$, we have

$$|\varepsilon' + 2\varepsilon_2|^2 + |\varepsilon''|^2 \to 0, \qquad (1.15)$$

which is satisfied when $\varepsilon' = -2\varepsilon_2$. Inserting it into the Drude model and taking $\varepsilon_2 = 1$ for air, we shall find out $\omega \cong \omega_p/\sqrt{3}$ for the dipolar resonance in the quasi-static approximation. One should keep in mind that the magnitude of α is also dependent on the nonzero imaginary part ε''.

So far, the polarizability in Eq. 1.14 holds for spherical particles. For an ellipsoidal particle with radii a, b, and c along x, y, and z axes, the polarizability along one of the axes α_i ($i = x, y, z$) can be written as

$$\alpha_i = 4\pi \frac{abc}{3} \frac{\varepsilon - \varepsilon_2}{\varepsilon_2 + L_i(\varepsilon - \varepsilon_2)}, \qquad (1.16)$$

in which L_i is the geometrical factor given by

$$L_i = \frac{abc}{2} \int_0^\infty \frac{dq}{(r_i^2 + q)\sqrt{(q+a^2)(q+b^2)(q+c^2)}}, \qquad (1.17)$$

where $r_i = a, b, c$. For spherical particles $a = b = c$, $L_i = 1/3$. Equation 1.16 is then simplified to Eq. 1.14.

As can be seen from Eq. 1.14, although the magnitude of the polarizability is dependent on the particle size, the resonance condition is not. The quasi-static approximation assumes that all electrons are driven by a homogeneous electric field and thus holds only for particles of a diameter much smaller than the wavelength. For large particles, the retardation effect becomes prominent, which leads to a red shift of the plasmon resonance. What is more, for large particles, the scattering process becomes prominent, so the radiation damping is also increased, leading to peak broadening. To include the retardation effect and radiation damping, the corrected polarizability for a nanosphere by the modified long-wavelength approximation (MLWA) [13] is used instead:

$$\alpha_{corr} = \frac{\alpha}{1 - \frac{1}{a}k^2 \frac{\alpha}{4\pi} - \frac{2}{3}ik^3 \frac{\alpha}{4\pi}}, \qquad (1.18)$$

in which k is the wave vector. The second term in the denominator corrects the retardation effect that results in the decreased resonance energy with increasing particle size, that is, red shift of the resonance peak, and the third term accounts for the radiative damping.

1.3.2 Extinction, Scattering, and Absorption

With the above-derived MLWA polarizability, the light scattering, absorption, and extinction cross sections of metal nanoparticles can be respectively calculated [14]:

$$C_{scat} = \frac{k^4}{6\pi}|\alpha|^2 = \frac{8}{3}\pi k^4 a^6 \left|\frac{\varepsilon - \varepsilon_2}{\varepsilon + 2\varepsilon_2}\right|^2, \qquad (1.19)$$

$$C_{abs} = k\operatorname{Im}\{\alpha\} = 4\pi k a^3 \operatorname{Im}\left\{\frac{\varepsilon - \varepsilon_2}{\varepsilon + 2\varepsilon_2}\right\}, \qquad (1.20)$$

and

$$C_{ext} = C_{scat} + C_{abs}, \qquad (1.21)$$

where $k = 2\pi/\lambda$. For small particles $a \ll \lambda$, the absorption efficiency is significantly larger than the scattering efficiency, so the absorption process dominates in the extinction. Since the scattering cross section scales as a^6, whereas the absorption cross section scales as a^3, the scattering process becomes more prominent when the size of the particle increases.

Figure 1.2a shows the scattering, absorption, and extinction cross sections of a gold nanoparticle of radius 40 nm in water, calculated using the MLWA polarizability and dielectric function of JC data. It can be seen that the absorption cross section at short wavelengths is much larger than the scattering cross section due to the interband absorption. Another feature is the slight deviation of peak positions in the scattering and absorption spectra. This can also be attributed to interband transitions. If one calculates the cross sections using the Drude dielectric function without considering interband transitions, the peak positions of scattering and absorption are the same. Figure 1.2b shows calculated extinction spectra of gold nanoparticles of radii 20 nm, 30 nm, and 40 nm in water and 20 nm in air using the same method as in Fig. 1.2a. The red shift of the spectrum for a = 20 nm in water with respect to that in air is due to the increase of the dielectric constant of the surrounding medium (see Eqs. 1.14 and 1.15), and the red shift with increasing size is explained by the retardation effect described in the preceding section.

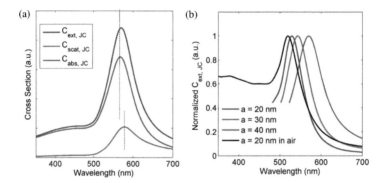

Figure 1.2 (a) Scattering, absorption, and extinction cross sections of a gold nanoparticle of radius 40 nm in water using the dielectric function of JC data. (b) Extinction cross sections of gold nanoparticles of radii 20 nm, 30 nm, and 40 nm in water and 20 nm in air.

1.3.3 Near-Field Distribution

The electric fields $\mathbf{E} = -\nabla \Phi$ inside and outside the particle can be calculated from the potentials in Eqs. 1.10 and 1.11 and written as [15]

$$\mathbf{E}_{in} = \frac{3\varepsilon_2}{\varepsilon + 2\varepsilon_2} \mathbf{E}_0, \quad (1.22)$$

and

$$\mathbf{E}_{out} = \mathbf{E}_0 + \frac{3\mathbf{n}(\mathbf{n} \cdot \mathbf{p}) - \mathbf{p}}{4\pi\varepsilon_0 \varepsilon_2} \frac{1}{r^3}, \quad (1.23)$$

where **n** is the unit vector, r is the distance from the dipole at the nanoparticle center, and **p** is the dipole moment defined by Eq. 1.13.

From Eq. 1.23, it is seen that the magnitude of the induced field decays as $1/r^3$ in the near field. The distribution of the induced electric field also follows that of the radiating dipole, that is, the fields are the strongest at the two poles along the orientation of the dipole. This is shown in Fig. 1.3, where a gold nanoparticle of 20 nm radius was excited by a linearly polarized plane wave at 520 nm in air. Note that the field enhancement $\left|\dfrac{\mathbf{E}}{\mathbf{E}_0}\right|$ is proportional to $\left|\dfrac{\varepsilon-\varepsilon_2}{\varepsilon+2\varepsilon_2}\right|$.

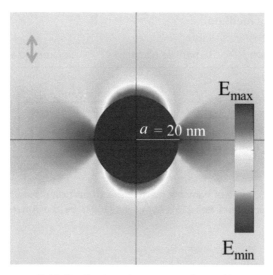

Figure 1.3 Near-field distribution of a 20 nm radius gold nanoparticle in air excited by a 520 nm plane wave. The arrow indicates the incident polarization direction.

1.3.4 Decay of Localized Surface Plasmons

The energy of localized SPs is damped via two mechanisms, radiative damping by emission of photons and nonradiative damping by creation of electron–hole pairs due to absorption (Fig. 1.4). The former occurs with the scattering event and, thus, is usually negligible for small particles ($a \ll \lambda$), but becomes prominent with increasing size of particles and dominates over the nonradiative damping if the particle size is comparable to the wavelength of incident light. The latter is mainly a result of electronic transitions. Depending on whether the excited electrons are in the conduction band or the d-band, the transition could be either intraband or interband. Through nonradiative decay, the energy of SPs is ultimately dissipated as heat. The total damping time (or dephasing time) is characterized by the linewidth of the particle plasmon resonance. In the frequency domain, full-width at half-maximum (FWHM, $\Delta\omega$) is related to the total dephasing time T through $\Delta\omega = 2/T$. T is typically of the order of femtoseconds (fs), for example, 5–10 fs for gold and silver nanoparticles, depending on the size and surrounding medium.

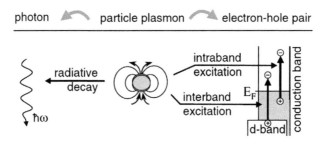

Figure 1.4 Scheme of damping mechanisms of particle plasmons. Reprinted with permission from Ref. [16]. Copyright (2002) by the American Physical Society.

1.4 Propagating Surface Plasmons

1.4.1 Propagating Surface Plasmons at the Planar Metal–Dielectric Interface

SPs can also be excited at the interface between a flat metal surface and a dielectric medium and propagate along the interface. The propagating SPs are usually called surface plasmon polaritons (SPPs). Solving Maxwell's equations under proper boundary conditions, one obtains the in-plane wave vector of SPPs at the planar interface between metal with relative permittivity $\varepsilon(\omega)$ and a dielectric medium with ε_2, as shown below:

$$k_{spp} = \frac{\omega}{c}\sqrt{\frac{\varepsilon(\omega)\varepsilon_2}{\varepsilon(\omega)+\varepsilon_2}}. \quad (1.24)$$

The resonance condition is found by minimizing the denominator: $\varepsilon + \varepsilon_2 \to 0$. At a low damping limit, substituting the real part of the Drude permittivity (Eq. 1.7) into the resonance condition, one finds

$$\omega_{sp} = \frac{\omega_p}{\sqrt{1+\varepsilon_2}}, \quad (1.25)$$

in which ω_p is the bulk plasmon frequency. If the dielectric medium is air, $\omega_{sp} = \omega_p/\sqrt{2}$.

In vacuum, the dispersion relation of light (the so-called light line) is written as $k_0 = \omega/c$, with c being the speed of light. In the medium with permittivity ε_2, the dispersion relation reads

$$k = \frac{\omega}{c}\sqrt{\varepsilon_2}. \quad (1.26)$$

The dispersion relation of SPPs (Eq. 1.24), together with the light line in medium with ε_2 and ε_{prism} (Eq. 1.26), is plotted in Fig. 1.5.

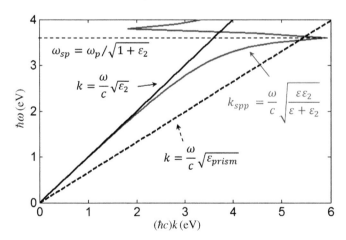

Figure 1.5 Dispersion relation of SPPs at the interface between Ag and a dielectric medium with permittivity $\varepsilon_2 = 1$.

From Fig. 1.5, it is clear that k_{spp} is always larger than the wave vector of light of the same energy in the dielectric medium with permittivity ε_2 for $\omega < \omega_{sp}$. This means that it is impossible to excite

SPPs directly using a plane wave from the medium. To match the wave vectors of light and SPPs, several methods have been employed [17], for example, prism coupling, grating coupling, and near-field coupling, as schemed in Fig. 1.6. For example, as shown in Fig. 1.6a, light is shined from a prism with a higher refractive index than that of the top medium—the Kretschmann configuration—so that the wave vector in the prism, k_{prism}, can be larger than k_{spp} (see Fig. 1.5) and its parallel component, $k_{prism,\|}$, can equal k_{spp}:

$$k_{prism,\|} = \frac{\omega}{c}\sqrt{\varepsilon_{prism}}\sin\theta = k_{spp}, \qquad (1.27)$$

where θ is the incidence angle. In practice, the reflection minimum occurs when the incident light is coupled to SPPs at incidence angle θ.

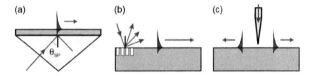

Figure 1.6 Excitation of SPPs using different methods: (a) prism coupling, (b) grating coupling, and (c) near-field coupling. Reprinted from Ref. [17], Copyright (2005), with permission from Elsevier.

It should be noted that in the Kretschmann configuration, it is the SPP at the top interface (between metal and dielectric of permittivity ε_2) that is excited. The fact that k_{spp} is larger than k in medium ε_2 indicates that the SPP does not radiate to this medium, although it leaks to the prism where $k_{prism} > k_{spp}$.

1.4.2 Length Scales of Propagating Surface Plasmons

As addressed in the preceding sections, the permittivity of metal in the dispersion relation of SPPs (Eq. 1.24) is a complex quantity in optical frequencies. Hence, k_{spp} can also be written in a complex form as

$$k_{spp} = k'_{spp} + ik''_{spp}. \qquad (1.28)$$

The first term in Eq. 1.28 describes the propagation of the SPPs, and the second term describes the damping due to absorption in the metal. Substituting the Drude model into Eq. 1.24, the real and imaginary parts of k_{spp} can be written as [18]

$$k'_{spp} = \frac{\omega}{c}\sqrt{\frac{\varepsilon'\varepsilon_2}{\varepsilon'+\varepsilon_2}}, \quad (1.29)$$

and

$$k''_{spp} = \frac{\omega}{c}\left(\frac{\varepsilon'\varepsilon_2}{\varepsilon'+\varepsilon_2}\right)^{3/2}\frac{\varepsilon''}{2\varepsilon'^2}, \quad (1.30)$$

where ε' and ε'' are the real part and the imaginary part of metal permittivity $\varepsilon(\omega)$, respectively. The SPP wavelength can be directly obtained from the real part of k_{spp} and reads

$$\lambda_{spp} = \frac{2\pi}{k'_{spp}} = \lambda_0 \left(\frac{\varepsilon'+\varepsilon_2}{\varepsilon'\varepsilon_2}\right)^{1/2}, \quad (1.31)$$

where λ_0 is the free-space wavelength. For the frequency smaller than the resonance frequency, the SPP wavelength is always shorter than the free-space wavelength (see Fig. 1.5).

The imaginary component in Eq. 1.28 suggests that as the SPP propagates, the amplitude decays exponentially as $\exp(-k''_{spp}x)$, and therefore the intensity decays as $\exp(-2k''_{spp}x)$, where x is the propagation distance. The propagation length of the SPPs, the distance over which the intensity falls to $1/e$ of its initial value, is written as

$$L_{spp} = \frac{1}{2k''_{spp}} = \lambda_0 \frac{\varepsilon'^2}{2\pi\varepsilon''}\left(\frac{\varepsilon'+\varepsilon_2}{\varepsilon'\varepsilon_2}\right)^{3/2}. \quad (1.32)$$

It is seen that a small imaginary part of $\varepsilon(\omega)$, that is, low-loss metal, is required to obtain a long propagation length. For noble metals, for example, Ag, L_{spp} for plasmons at the Ag–air interface can reach ~100 μm if excited by light of 1000 nm wavelength, neglecting radiative damping [18]. In reality, the propagation length is typically up to tens of microns in the visible frequencies.

As mentioned earlier (Eq. 1.26), the wave vector of light in the medium of ε_2 is $\sqrt{\varepsilon_2}\,k_0$. The relationship between the light wave vector and the in-plane wave vector of SPPs and the z component (normal to the interface) of the wave vector in the medium is as follows:

$$\varepsilon_2 k_0^2 = k_{spp}^2 + k_z^2. \quad (1.33)$$

Since $k_{spp} > \sqrt{\varepsilon_2}\, k_0$, k_z should be imaginary in the dielectric medium. Similarly, k_z is also imaginary in the metal [18]. This means that the fields should decay exponentially with distance away from the interface. The penetration depths in the dielectric medium (δ_2) and metal (δ_m)—where the amplitude falls to $1/e$ of the initial value—can be found as

$$\delta_2 = \frac{1}{k_0}\left|\frac{\varepsilon' + \varepsilon_2}{\varepsilon_2^2}\right|^{1/2}, \qquad (1.34)$$

and

$$\delta_m = \frac{1}{k_0}\left|\frac{\varepsilon' + \varepsilon_2}{\varepsilon'^2}\right|^{1/2}. \qquad (1.35)$$

As the absolute value of the real part of metal permittivity is larger than the permittivity of the medium in optical frequencies, the penetration depth in the medium is longer than in the metal by a factor of $|\varepsilon'/\varepsilon_2|$. In the visible region for noble metals, δ_m is typically ~25–30 nm and δ_2 can be approximately half of the light wavelength.

The two length scales, that is, propagation length (L_{spp} that is of the order of micrometers) and penetration depth (δ_m and δ_2 that are of tens to hundreds of nanometers), define the most important characteristics of SPPs—long-distance propagation and subwavelength confinement of EM energy.

1.5 Research Topics

The research in both the fundamentals and the applications of plasmonics has seen prosperous growth in recent years. Figure 1.7 shows an overview of the main research topics and potential applications in plasmonics. Most studies are based on surface plasmon resonances (SPRs) and localized surface plasmon resonances (LSPRs), plasmon coupling between nanostructures, and waveguiding at the subwavelength scale.

The SPRs/LSPRs are strongly dependent on the nanostructure geometries and environments. The sensitive response of the resonance frequency to environmental change is attractive for chemical and biological sensing applications [1, 19]. Plasmonic rulers have been developed on the basis of the sensitive dependence of the LSPR wavelength on the distance between two coupling metal nanoparticles [20, 21]. The LSPRs are also the basis for the large field enhancement and tight field confinement in plasmonic nanostructures. Moreover, SPRs/LSPRs are employed for sub-diffraction-limited optical imaging [22, 23] and nanolithography [24, 25]. The absorption induced by SPRs/LSPRs can result in heat generation, which has been used for cancer therapy [26], heat-assisted magnetic recording [27], solar vapor generation [28], and photodetection [29]. SPRs/LSPRs also play critical roles in the enhanced optical transmission of nanoholes in metal films [30, 31].

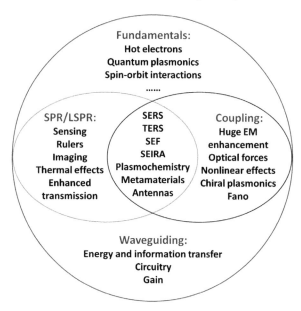

Figure 1.7 Summary of the most prominent SP characteristics and research topics in the plasmonics field.

The plasmon coupling between adjacent metal nanostructures can largely shift the LSPR frequencies and generate hugely enhanced EM fields in the nanogaps between the nanostructures. The huge field enhancement is probably the most prominent property of SPs for a large variety of applications. Plasmon-enhanced spectroscopy based on this field enhancement effect has been widely investigated, such as surface-enhanced Raman scattering (SERS) [32, 33], tip-enhanced Raman scattering (TERS) [34], surface-enhanced fluorescence (SEF) [35–37], and surface-enhanced infrared absorption (SEIRA) [38], which add new tools for ultrasensitive detection and imaging. Plasmonic field enhancement has also been used to amplify some nonlinear effects, such as white-light supercontinuum generation [39], four-wave mixing [40], harmonic generation [41], lasing [42], and coherent anti-Stokes Raman scattering [43]. Optical forces can be enhanced by the concentrated field in coupled plasmonic nanostructures as well [44, 45]. On the basis of SPRs/LSPRs and plasmon coupling, some other topics have also been developed, including plasmochemistry [46, 47], plasmonic metamaterials and metasurfaces [48, 49], nanoantennas [50–52], chiral plasmonics [53], and Fano resonances [54].

SPs can propagate in 1D metal nanostructures with tight field confinement beyond the diffraction limit, which enables the construction of nanophotonic circuits with small footprints for information technologies. Quite a few different 1D plasmonic waveguides have been investigated, such as nanowires [55] and nanoslots/nanogrooves in metal films [56, 57]. Due to the plasmon coupling between neighbored nanoparticles, a metal nanoparticle chain can also support SP propagation over a short distance [58, 59]. The Ohmic loss in metal nanostructures leads to the SP damping during propagation. To overcome this problem, some schemes using gain materials to compensate the SP loss can be employed [60–62]. The combination of metal nanostructures and gain materials also enables the development of some active plasmonic devices [63]. Besides nanowaveguides, various components are needed to build plasmonic circuitry, including nanolasers, photodetectors, modulators, routers, logic gates, etc. The integration of different components into functional circuitry is being pursued and still a highly challenging topic [64].

The studies about the fundamentals of SPs have made great progress in recent years. Plasmon-generated hot electrons attract much attention, as they may play important roles in photochemical and photovoltaic processes [65]. But more effort is needed to reveal how the hot electrons are involved in those processes. Quantum plasmonics is another hot topic, which mainly includes studies on the quantum treatments of the theoretical descriptions for SPs and the quantum properties of SPs. When the size of metal nanoparticles decreases to a few nanometers, or the distance between two coupling nanostructures becomes less than 1 nm, the nonlocality of the metal's dielectric response and the electron tunneling will make the SPR frequency and the field enhancement largely deviate from the values predicted by the classical electrodynamics [66–70]. A quantum-corrected model is developed to deal with the coupling across subnanometer gaps [71]. The quantum properties of single quantized SPs, for example, wave-particle duality, entanglement, and quantum interference, are investigated in the context of quantum nanophotonic circuits [72–75], which can also be assigned to the waveguiding and circuitry part in Fig. 1.7. These studies combine plasmonics and quantum optics and open the prospects of using quantum optical techniques to investigate and manipulate single SPs. The spin–orbit interactions of light can be largely enhanced in plasmonic nanostructures. This effect can be explored for novel spin–orbit optical phenomena and new functionalities in nanophotonic devices and circuits by coupling the spin and orbital degrees of freedom of photons [76, 77]. The research on this topic is attracting growing interest. In recent years, investigations on new plasmonic materials, for example, graphene [78], have become prosperous. The development of graphene plasmonics pushes the frequency of plasmons from visible and near infrared to mid infrared and expands the realms of plasmonics.

References

1. K. M. Mayer and J. H. Hafner, *Chem. Rev.*, **111**, 3828–3857 (2011).
2. W. L. Barnes, A. Dereux, and T. W. Ebbesen, *Nature*, **424**, 824–830 (2003).
3. M. Moskovits, *J. Raman Spectrosc.*, **36**, 485–496 (2005).
4. E. Hutter and J. H. Fendler, *Adv. Mater.*, **16**, 1685–1706 (2004).
5. K. A. Willets and R. P. Van Duyne, *Annu. Rev. Phys. Chem.*, **58**, 267–297 (2007).
6. S. Gwo and C. K. Shih, *Rep. Prog. Phys.*, **79**, 086501 (2016).
7. H. A. Atwater and A. Polman, *Nat. Mater.*, **9**, 205–213 (2010).
8. E. Ozbay, *Science*, **311**, 189–193 (2006).
9. A. Polman, *Science*, **322**, 868–869 (2008).
10. M. Faraday, *Philos. Trans. R. Soc. London*, **147**, 145 (1857).
11. G. Mie, *Ann. Phys.*, **25**, 377–445 (1908).
12. P. B. Johnson and R. W. Christy, *Phys. Rev. B*, **6**, 4370–4379 (1972).
13. M. Meier and A. Wokaun, *Opt. Lett.*, **8**, 581–583 (1983).
14. C. F. Bohren and D. R. Huffman, *Absorption and Scattering of Light by Small Particles* (John Wiley & Sons, New York, 1983).

15. S. A. Maier, *Plasmonics: Fundamentals and Applications* (Springer, 2006).
16. C. Sönnichsen, T. Franzl, T. Wilk, G. von Plessen, J. Feldmann, O. Wilson, and P. Mulvaney, *Phys. Rev. Lett.*, **88**, 077402 (2002).
17. A. V. Zayats, Smolyaninov, II, and A. A. Maradudin, *Phys. Rep.*, **408**, 131–314 (2005).
18. W. L. Barnes, *J. Opt. A: Pure Appl. Opt.*, **8**, S87–S93 (2006).
19. B. Liedberg, C. Nylander, and I. Lundstrom, *Sens. Actuators*, **4**, 299–304 (1983).
20. C. Sonnichsen, B. M. Reinhard, J. Liphardt, and A. P. Alivisatos, *Nat. Biotechnol.*, **23**, 741–745 (2005).
21. P. K. Jain, W. Y. Huang, and M. A. El-Sayed, *Nano Lett.*, **7**, 2080–2088 (2007).
22. J. B. Pendry, *Phys. Rev. Lett.*, **85**, 3966–3969 (2000).
23. X. Zhang and Z. W. Liu, *Nat. Mater.*, **7**, 435–441 (2008).
24. X. G. Luo and T. Ishihara, *Appl. Phys. Lett.*, **84**, 4780–4782 (2004).
25. Z. W. Liu, Q. H. Wei, and X. Zhang, *Nano Lett.*, **5**, 957–961 (2005).
26. R. Bardhan, S. Lal, A. Joshi, and N. J. Halas, *Acc. Chem. Res.*, **44**, 936–946 (2011).
27. W. A. Challener, C. B. Peng, A. V. Itagi, D. Karns, W. Peng, Y. Y. Peng, X. M. Yang, X. B. Zhu, N. J. Gokemeijer, Y. T. Hsia, G. Ju, R. E. Rottmayer, M. A. Seigler, and E. C. Gage, *Nat. Photonics*, **3**, 220–224 (2009).
28. O. Neumann, A. S. Urban, J. Day, S. Lal, P. Nordlander, and N. J. Halas, *ACS Nano*, **7**, 42–49 (2013).
29. W. Zhang, Z. P. Li, Z. Q. Guan, H. Shen, W. B. Yu, W. D. He, X. P. Yan, P. Li, and H. X. Xu, *Chin. Sci. Bull.*, **57**, 68–71 (2012).
30. T. W. Ebbesen, H. J. Lezec, H. F. Ghaemi, T. Thio, and P. A. Wolff, *Nature* **391**, 667–669 (1998).
31. F. J. Garcia-Vidal, L. Martin-Moreno, T. W. Ebbesen, and L. Kuipers, *Rev. Mod. Phys.*, **82**, 729–787 (2010).
32. H. X. Xu, E. J. Bjerneld, M. Käll, and L. Borjesson, *Phys. Rev. Lett.*, **83**, 4357–4360 (1999).
33. H. X. Xu, J. Aizpurua, M. Käll, and P. Apell, *Phys. Rev. E*, **62**, 4318–4324 (2000).
34. B. Pettinger, P. Schambach, C. J. Villagomez, and N. Scott, *Annu. Rev. Phys. Chem.*, **63**, 379–399 (2012).
35. P. Anger, P. Bharadwaj, and L. Novotny, *Phys. Rev. Lett.*, **96**, 113002 (2006).

36. S. Kuhn, U. Hakanson, L. Rogobete, and V. Sandoghdar, *Phys. Rev. Lett.*, **97**, 017402 (2006).
37. J. Zhang, Y. Fu, M. H. Chowdhury, and J. R. Lakowicz, *Nano Lett.*, **7**, 2101–2107 (2007).
38. L. V. Brown, K. Zhao, N. King, H. Sobhani, P. Nordlander, and N. J. Halas, *J. Am. Chem. Soc.*, **135**, 3688–3695 (2013).
39. P. Muhlschlegel, H. J. Eisler, O. J. F. Martin, B. Hecht, and D. W. Pohl, *Science,* **308**, 1607–1609 (2005).
40. M. Danckwerts and L. Novotny, *Phys. Rev. Lett.*, **98**, 026104 (2007).
41. S. Kim, J. H. Jin, Y. J. Kim, I. Y. Park, Y. Kim, and S. W. Kim*, Nature,* **453**, 757–760 (2008).
42. J. Y. Suh, C. H. Kim, W. Zhou, M. D. Huntington, D. T. Co, M. R. Wasielewski, and T. W. Odom, *Nano Lett.*, **12**, 5769–5774 (2012).
43. Y. Zhang, Y. R. Zhen, O. Neumann, J. K. Day, P. Nordlander, and N. J. Halas, *Nat. Commun.*, **5**, 4424 (2014).
44. H. X. Xu and M. Käll, *Phys. Rev. Lett.*, **89**, 246802 (2002).
45. M. L. Juan, M. Righini, and R. Quidant*, Nat. Photonics,* **5**, 349–356 (2011).
46. M. T. Sun, Z. L. Zhang, H. R. Zheng, and H. X. Xu, *Sci. Rep.*, **2**, 647 (2012).
47. S. Linic, P. Christopher, and D. B. Ingram, *Nat. Mater.*, **10**, 911–921 (2011).
48. K. Yao and Y. M. Liu, *Nanotechnol. Rev.*, **3**, 177–210 (2014).
49. N. F. Yu and F. Capasso, *Nat. Mater.*, **13**, 139–150 (2014).
50. T. Shegai, Z. P. Li, T. Dadosh, Z. Y. Zhang, H. X. Xu, and G. Haran, *Proc. Natl. Acad. Sci. USA*, **105**, 16448–16453 (2008).
51. Z. P. Li, T. Shegai, G. Haran, and H. X. Xu, *ACS Nano*, **3**, 637–642 (2009).
52. L. Novotny and N. van Hulst, *Nat. Photonics*, **5**, 83–90 (2011).
53. A. Guerrero-Martinez, J. L. Alonso-Gomez, B. Auguie, M. M. Cid, and L. M. Liz-Marzan, *Nano Today*, **6**, 381–400 (2011).
54. B. Luk'yanchuk, N. I. Zheludev, S. A. Maier, N. J. Halas, P. Nordlander, H. Giessen, and C. T. Chong, *Nat. Mater.*, **9**, 707–715 (2010).
55. H. Wei and H. X. Xu, *Nanophotonics*, **1**, 155–169 (2012).
56. G. Veronis and S. H. Fan, *Opt. Lett.*, **30**, 3359–3361 (2005).
57. S. I. Bozhevolnyi, V. S. Volkov, E. Devaux, J. Y. Laluet, and T. W. Ebbesen*, Nature,* **440**, 508–511 (2006).
58. S. A. Maier, P. G. Kik, H. A. Atwater, S. Meltzer, E. Harel, B. E. Koel, and A. A. G. Requicha, *Nat. Mater.*, **2**, 229–232 (2003).

59. Z. P. Li, and H. X. Xu, *J. Quant. Spectrosc. Radiat. Transfer*, **103**, 394–401 (2007).
60. I. De Leon and P. Berini, *Nat. Photonics* **4**, 382–387 (2010).
61. M. C. Gather, K. Meerholz, N. Danz, and K. Leosson, *Nat. Photonics,* **4**, 457–461 (2010).
62. N. Liu, H. Wei, J. Li, Z. X. Wang, X. R. Tian, A. L. Pan, and H. X. Xu, *Sci. Rep.*, **3**, 1967 (2013).
63. O. Hess, J. B. Pendry, S. A. Maier, R. F. Oulton, J. M. Hamm, and K. L. Tsakmakidis, *Nat. Mater.*, **11**, 573–584 (2012).
64. K. C. Y. Huang, M. K. Seo, T. Sarmiento, Y. J. Huo, J. S. Harris, and M. L. Brongersma, *Nat. Photonics,* **8**, 244–249 (2014).
65. M. L. Brongersma, N. J. Halas, and P. Nordlander, *Nat. Nanotechnol.*, **10**, 25–34 (2015).
66. L. Mao, Z. P. Li, B. Wu, and H. X. Xu, *Appl. Phys. Lett.*, **94**, 243102 (2009).
67. J. Zuloaga, E. Prodan, and P. Nordlander, *Nano Lett.*, **9**, 887–891 (2009).
68. K. J. Savage, M. M. Hawkeye, R. Esteban, A. G. Borisov, J. Aizpurua, and J. J. Baumberg, *Nature,* **491**, 574–577 (2012).
69. C. Ciraci, R. T. Hill, J. J. Mock, Y. Urzhumov, A. I. Fernandez-Dominguez, S. A. Maier, J. B. Pendry, A. Chilkoti, and D. R. Smith, *Science,* **337**, 1072–1074 (2012).
70. J. A. Scholl, A. L. Koh, and J. A. Dionne, *Nature,* **483**, 421–427 (2012).
71. R. Esteban, A. G. Borisov, P. Nordlander, and J. Aizpurua, *Nat. Commun.*, **3**, 825 (2012).
72. A. V. Akimov, A. Mukherjee, C. L. Yu, D. E. Chang, A. S. Zibrov, P. R. Hemmer, H. Park, and M. D. Lukin, *Nature,* **450**, 402–406 (2007).
73. R. Kolesov, B. Grotz, G. Balasubramanian, R. J. Stohr, A. A. L. Nicolet, P. R. Hemmer, F. Jelezko, and J. Wrachtrup, *Nat. Phys.*, **5**, 470–474 (2009).
74. A. Gonzalez-Tudela, D. Martin-Cano, E. Moreno, L. Martin-Moreno, C. Tejedor, and F. J. Garcia-Vidal, *Phys. Rev. Lett.*, **106**, 020501 (2011).
75. R. W. Heeres, L. P. Kouwenhoven, and V. Zwiller, *Nat. Nanotechnol.*, **8**, 719–722 (2013).
76. K. Y. Bliokh, F. J. Rodriguez-Fortuno, F. Nori, and A. V. Zayats, *Nat. Photonics,* **9**, 796–808 (2015).
77. D. Pan, H. Wei, L. Gao, and H. X. Xu, *Phys. Rev. Lett.*, **117**, 166803 (2016).
78. F. J. G. de Abajo, *ACS Photonics*, **1**, 135–152 (2014).

Chapter 2

Light Scattering by Small Metallic Particles: Mie Theory and More

Shunping Zhang[a] and Hongxing Xu[a,b]

[a]*School of Physics and Technology, Wuhan University, Wuhan 430072, China*
[b]*Institute for Advanced Studies, Wuhan University, Wuhan 430072, China*
spzhang@whu.edu.cn; hxxu@whu.edu.cn

Light scattering by small particles raised considerable interest in ancient times, usually associated with coloring. In the current era, it serves as basic knowledge for both astronomy and atmospheric science. Particularly, light scattering by small particles (10–500 nm) attract renewed scientific interest due to the rapid growing of nanophotonics. When a photon collides with an object, it interacts with the object and ends up with a change in its own state, for example, the energy (wavelength), linear momentum (direction), and polarization (angular momentum). Photons can also be created or annihilated during the interaction, leading to a change in the number state of the photons. These changes in the initial and final states of the photons carry the information of the object. Therefore, analyzing the probing beam offers a powerful strategy looking into the target object. For example, when an X-ray beam is traveling within a crystal, it's strongly diffracted by the atom arrays since its wavelength

---

Nanophotonics: Manipulating Light with Plasmons
Edited by Hongxing Xu
Copyright © 2018 Pan Stanford Publishing Pte. Ltd.
ISBN 978-981-4774-14-7 (Hardcover), 978-1-315-19661-9 (eBook)
www.panstanford.com

is of the order of the lattice constant of the crystal. By analyzing the diffracted beam, we can know about the crystalline structures of the sample. At the long-wavelength end of the electromagnetic (EM) spectrum, the scattering of a radio wave can be used in radar detection for remote sensing of the macroscale objects. Light scattering discussed in this chapter usually (but not restricted to) refers to the scattering of the ultraviolet-visible-infrared EM waves by nano-/microscale objects ranging from single atoms to small particles.

Light scattering can be more specifically divided into different categories. Depending on whether the energy of the photon is changed or not, the scattering process can be inelastic or elastic. The Raman/Brillouin scattering process is typical inelastic light scattering whose intensity is usually orders of magnitude smaller than the elastic scattering. In this chapter, we restrict our discussions within the frame of elastic light scattering by small nanoparticles (NPs), in which the material response is modeled by an isotropic dielectric permittivity and permeability. Elastic light scattering, however, can be further divided into three regions according to a size parameter x defined as the ratio of the size of the object and the wavelength, as shown in Fig. 2.1. For $x \ll 1$, the object can be viewed as a point dipole characterized by a polarizability tensor. This is called Rayleigh scattering in honor of Lord Rayleigh for his pioneer works in 1899 [1]. On the contrary, geometric optics works when the condition $x \gg 1$ holds. Tracing the reflection/refraction events and the light paths is enough to know about the properties of the optical systems. For the intermediate region, $x \approx 1$ (the size of the object is comparable to the wavelength), a full vectorial description of the EM field is required. For a spherical object, this region corresponds to Mie scattering.

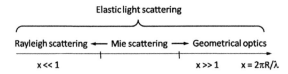

Figure 2.1 Elastic light scattering in different regimes. R is the (effective) radius of the particle, and λ is the wavelength of light.

To solve a full-wave EM problem, one has to solve Maxwell's equations for given boundary conditions. For a symmetric object, such as a sphere or an infinitely long cylinder, analytical solutions are available. However, for irregularly shaped scatterers, numerical techniques are required to discretize the object into small pieces, either over the entire volume or over the object surface. In this chapter, we will introduce in Section 2.1 the scattering of EM waves by a single sphere—Mie theory. Generalized Mie theory for concentric spheres or multiple spheres will be introduced in Section 2.2. Numerical techniques for nonspherical particles will be briefly summarized in Section 2.3.

2.1 Light Scattering by a Single Spherical Particle and Mie Theory

The scattering of a monochromatic plane wave by a spherical particle was first deduced by Gustav Mie in 1908, by introducing scalar EM potentials and applying the boundary value conditions at the spherical interface [2]. Using a similar method, Bohn greatly reduced the lengthy derivation by Mie in his famous book [3]. A complete but compact form of Mie theory, regarded as conventional Mie theory, was first proposed by Stratton in 1941 [4], where vector spherical harmonics (VSH) were introduced. Later, C. F. Bohren and D. R. Huffman (1983) discussed the scattering of light by small particles within the frame of Mie theory in detail [5]. In fact, Lorenz had considered the same problem earlier than Mie, in 1890, and therefore the term "Mie theory" can also be called "Lorenz–Mie theory." In this section, we will introduce the vector solutions to the vector wave equation, following the discussions by Bohn and Stratton. The expansion of a plane wave into VSH and conventional Mie theory will be presented right after that.

2.1.1 Vector Solutions in the Spherical Coordinate: VSH

The analytical solutions to the scalar Helmholtz equation $(\nabla^2 + k^2)\phi(\mathbf{r}) = 0$ in the spherical coordinate take the form of

$$\phi_{nm}(r,\theta,\varphi) = z_n^{(j)}(kr) Y_n^m(\theta,\varphi), |m| \le n, 0 \le n \le \infty, \quad (2.1)$$

where

$$Y_n^m(\theta,\varphi) = (-1)^m \sqrt{\frac{2n+1}{4\pi} \cdot \frac{(n-m)!}{(n+m)!}} P_n^m(\cos\theta) e^{im\varphi} \quad (2.2)$$

is spherical harmonics and $P_n^m(\cos\theta)$ the associated Legendre functions. $z_n^{(j)}(kr)$ represents the spherical Bessel function $j_n(kr)$, the Neumann function $y_n(kr)$, and the Hankel function of the first kind $h_n^{(1)}(kr)$ and the second kind $h_n^{(2)}(kr)$ for j = 1, 2, 3, and 4, respectively. According to the completeness of the eigenfunction set$\{\phi_{nm}\}$, any function that satisfies the scalar wave equation could be expanded as an infinite series in ϕ_{nm}.

Similarly, the solutions to the vector differential wave equation $(\nabla^2 + k^2)\mathbf{F}(\mathbf{r}) = 0$ define a complete set of VSH, which actually can be generated from scalar spherical wave functions by the use of certain operators:

$$\mathbf{L}_{nm} \equiv \nabla \phi_{nm}(\mathbf{r})$$
$$\mathbf{M}_{nm} \equiv (-\mathbf{r} \times \nabla)\phi_{nm}(\mathbf{r}) \qquad (2.3)$$
$$\mathbf{N}_{nm} \equiv k^{-1}\nabla \times \mathbf{M}_{nm}$$

The angular momentum operator $\hat{L} = -\mathbf{r} \times \nabla$ defined here is slightly different from that frequently used in quantum mechanics by an imaginary unit i. Knowing that $\mathbf{M}_{nm} = \nabla \times [\mathbf{r}\phi_{nm}(\mathbf{r})]$ and the divergence of the curl of any vector function vanishes and vice versa, we find

$$\nabla \cdot \mathbf{M}_{nm} = 0, \quad \nabla \cdot \mathbf{N}_{nm} = 0, \quad \nabla \times \mathbf{L}_{nm} = 0 \qquad (2.4)$$

immediately. The operator $\hat{L}$ is commutative with the Laplace operator by $\hat{L}\nabla^2 = \nabla^2 \hat{L}$. Note that $\phi_{nm}(\mathbf{r})$ satisfies the scalar wave equation. We, therefore, have $\mathbf{M}_{nm}$ satisfying the vector Helmholtz equation:

$$\nabla^2 \mathbf{M}_{nm} + k^2 \mathbf{M}_{nm} = 0. \qquad (2.5)$$

By using vector identities and their definition, $\mathbf{N}_{nm}$ can be proved to satisfy the vector Helmholtz equation as well. By definition, $\mathbf{M}_{nm}$ and $\mathbf{N}_{nm}$ are correlated by

$$k\mathbf{N}_{nm} = \nabla \times \mathbf{M}_{nm}$$
$$k\mathbf{M}_{nm} = \nabla \times \mathbf{N}_{nm} \qquad (2.6)$$

in analogy to the electric field $\mathbf{E}$ and magnetic field $\mathbf{H}$ of an EM wave. Therefore, each VSH field represents a special partial field. Note that no radial component of $\mathbf{M}_{nm}$ exists. Therefore, for E modes (transverse magnetic [TM]), $\mathbf{M}_{nm}$ represents the magnetic field and

$\mathbf{N}_{nm}$ the electric field. For H modes (transverse electric [TE]), the opposite is true. However, an **L**-type field is a purely longitudinal wave. Therefore, for a given purely solenoidal field ($\nabla \times \mathbf{F} \neq 0$), only **M**- and **N**-type fields are necessary for expansion, while for an irrotational field, terms in **L** are sufficient to provide an expansion. Since the field concerned in the chapter is divergence free, an **L**-type field is omitted in the following discussions. For convenience, the VSH are written in matrix form

$$\mathbf{M}_{nm} = (\hat{r}, \hat{\theta}, \hat{\varphi}) \begin{bmatrix} 0 \\ \dfrac{1}{\sin\theta} \dfrac{\partial}{\partial \varphi} \\ -\dfrac{\partial}{\partial \theta} \end{bmatrix} z_n^{(j)}(kr) Y_n^m(\theta, \varphi)$$

$$\mathbf{N}_{nm} = (\hat{r}, \hat{\theta}, \hat{\varphi}) \begin{bmatrix} n(n+1) z_n^{(j)}(kr) \\ \dfrac{\partial}{\partial r}\{z_n^{(j)}(kr)\} \dfrac{\partial}{\partial \theta} \\ \dfrac{\partial}{\partial r}\{rz_n^{(j)}(kr)\} \dfrac{1}{\sin\theta} \dfrac{\partial}{\partial \varphi} \end{bmatrix} \dfrac{1}{kr} Y_n^m(\theta, \varphi)$$

(2.7)

The concept of state in quantum mechanics can also be borrowed to describe the VSH and the EM fields. This does not mean that quantum physics is used to solve EM problems. However, it is convenient to use the concept of state, which also captures the nature of EM problems. Here, Dirac notation and rules are used (see Sakurai [6]) for the VSH. $|\,\rangle$ denotes the ket, while $\langle\,|$ denotes the bra. Then, Eq. 2.7 can be written as $|nmjp\rangle$ with $p = 1$ for **M** and with $p = 2$ for **N**, where j represents different spherical Bessel functions.

2.1.2 Expansion of a Plane Wave by VSH

For a linearly polarized plane wave, the electric field $|i, E\rangle = \mathbf{E}_0 e^{i\mathbf{k}\cdot\mathbf{r}}$ can be expanded in an infinite series of VSH as

$$|i, E\rangle = \sum_{n=1}^{\infty} \sum_{m=-n}^{n} C1_{mn} |nm11\rangle + C2_{mn} |nm12\rangle, \qquad (2.8)$$

where the expansion coefficients can be determined concisely by $C1_{mn} = \frac{\langle nm11|i\rangle}{\langle nm11|nm11\rangle}$ and $C2_{mn} = \frac{\langle nm12|i\rangle}{\langle nm12|nm12\rangle}$. The integrals run over all solid angles. Noting the orthogonal characteristics of the VSH, the expansion coefficients take the following form [7]:

$$C1_{mn} = \frac{2\pi i^n}{n(n+1)} \Big\{ (iE_x - E_y)\sqrt{(n+m+1)(n-m)} Y_n^{m+1*}(\hat{\mathbf{k}})$$

$$+ (iE_x + E_y)\sqrt{(n-m+1)(n+m)} Y_n^{m-1*}(\hat{\mathbf{k}}) + 2imE_z Y_n^{m*}(\hat{\mathbf{k}}) \Big\} \quad (2.9)$$

and

$$C2_{mn} = \frac{2\pi i^n}{n(n+1)} \Bigg\{ -n\sqrt{\frac{(n+m+1)(n+m+2)}{(2n+1)(2n+3)}} (iE_x - E_y) Y_{n+1}^{m+1*}(\hat{\mathbf{k}})$$

$$+ n\sqrt{\frac{(n-m+1)(n-m+2)}{(2n+1)(2n+3)}} (iE_x + E_y) Y_{n+1}^{m-1*}(\hat{\mathbf{k}})$$

$$+ (n+1)\sqrt{\frac{(n-m-1)(n-m)}{(2n-1)(2n+1)}} (-iE_x + E_y) Y_{n-1}^{m+1*}(\hat{\mathbf{k}}) \quad (2.10)$$

$$+ (n+1)\sqrt{\frac{(n+m-1)(n+m)}{(2n-1)(2n+1)}} (iE_x + E_y) Y_{n-1}^{m-1*}(\hat{\mathbf{k}})$$

$$+ 2in\sqrt{\frac{(n+m+1)(n-m+1)}{(2n+1)(2n+3)}} E_z Y_{n+1}^{m*}(\hat{\mathbf{k}}) - 2i(n+1)$$

$$\sqrt{\frac{(n+m)(n-m)}{(2n-1)(2n+1)}} E_z Y_{n-1}^{m*}(\hat{\mathbf{k}}) \Bigg\},$$

where $\hat{\mathbf{k}}$ denotes the unit vector of the incident direction.

The expansion of the magnetic field is easily obtained by applying the relation between the electric field **E** and the magnetic field **H**, $\mathbf{H} = \frac{1}{i\omega\mu} \nabla \times \mathbf{E}$, and the relations between the normal modes (Eq. 2.6). The incident magnetic field is thus expanded as

$$|i,H\rangle = \frac{k}{i\omega\mu} (C1_{mn}|nm12\rangle + C2_{mn}|nm11\rangle). \quad (2.11)$$

Here, the summation signs are omitted, which are the same as in Eq. 2.8, for convenience. The summation over n and m is assumed in the following section as well.

2.1.3 Scattering by a Single Sphere

Conventional Mie theory deals with the scattering of an EM wave by a sphere with an arbitrary size in a homogeneous medium, given that the material response can be described by an isotropic scalar dielectric function. Suppose there's no net charge or current in the space and the field varies harmonically with time via exp($-i\omega t$). Maxwell's equations take the form of

$$\begin{aligned} \nabla \times \mathbf{E} &= i\omega\mu\mathbf{H} \\ \nabla \times \mathbf{H} &= -i\omega\varepsilon\mathbf{E} \\ \nabla \cdot \mathbf{E} &= 0 \\ \nabla \cdot \mathbf{H} &= 0 \end{aligned} \quad , \tag{2.12}$$

where $\mu = \mu_r\mu_0$ and $\varepsilon = \varepsilon_r\varepsilon_0$ (μ_0 and ε_0 are permeability and permittivity in vacuum) are local permeability and permittivity, respectively. Using the vector identity $\nabla \times \nabla \times \mathbf{A} = \nabla(\nabla \cdot \mathbf{A}) - \nabla^2\mathbf{A}$, we immediately obtain the Helmholtz equations

$$\begin{cases} (\nabla^2 + k^2)\mathbf{E}(\mathbf{r}) = 0 \\ (\nabla^2 + k^2)\mathbf{H}(\mathbf{r}) = 0 \end{cases} , \tag{2.13}$$

where $k = \omega\sqrt{\mu\varepsilon}$ defines the magnitude of the wave vector inside or outside the sphere. For nonmagnetic materials, the relative permeability is unity everywhere ($\mu_r = 1$). For conductive materials, the permittivity becomes complex and is related to the electrical conductivity σ by $\text{Im}(\tilde{\varepsilon}) = \sigma/\omega$. Therefore, the refractive index of the conductive materials becomes complex according to $\tilde{n}^2 = \mu_r\tilde{\varepsilon}_r$.

From Section 2.1.1, we know that the solution to Eq. 2.13 can be expanded into a series of VSH. Noting that the outward-going scattered wave should vanish at a distance far from the scatterer by physical intuition, we choose $h_n^{(1)}(x)$ for the scattered field outside the sphere. For fields inside the sphere, $j_n(x)$ is proper for the sake of avoiding singularity at the center of the sphere. Thus, the scattered and transmitted fields can be expanded as

$$\begin{aligned} |s, E\rangle &= C1_{mn} b_n |nm31\rangle + C2_{mn} a_n |nm32\rangle \\ |s, H\rangle &= \frac{k}{i\omega\mu}(C1_{mn} b_n |nm32\rangle + C2_{mn} a_n |nm31\rangle) \\ |t, E\rangle &= C1_{mn} d_n |nm11\rangle + C2_{mn} c_n |nm12\rangle \\ |t, H\rangle &= \frac{k}{i\omega\mu}(C1_{mn} d_n |nm12\rangle + C2_{mn} c_n |nm11\rangle) \end{aligned} , \tag{2.14}$$

in which the unknown coefficients a_n, b_n, c_n, and d_n are the so-called Mie coefficients in the literature. The total EM field outside the sphere is the sum of the incident field and scattered field, as

$$\mathbf{E}_{\text{total}} = \mathbf{E}_i + \mathbf{E}_s. \qquad (2.15)$$

To determine the coefficients, the boundary conditions are imposed on the surface of the sphere. It requires that the tangent components of the electric and magnetic fields be continuous at the interface, as

$$\begin{aligned}\mathbf{n} \times \mathbf{E}_{\text{total}}\big|_{r=R} &= \mathbf{n} \times \mathbf{E}_t\big|_{r=R} \\ \mathbf{n} \times \mathbf{H}_{\text{total}}\big|_{r=R} &= \mathbf{n} \times \mathbf{H}_t\big|_{r=R}\end{aligned}, \qquad (2.16)$$

where $\mathbf{n} = \hat{\mathbf{r}}$ is the unit normal vector of the sphere surface. As the electric and magnetic fields are expanded into infinite series of VSH, Eq. 2.16 actually contains $4N \times (2N+1)$ ($N \to \infty$ is the maximum of n) unknown coefficients corresponding to different n and m. However, if we coordinate the system so that the z axis is parallel to the incident direction and the x axis is parallel to the electric field, the expansion coefficients $C1_{nm}$ and $C2_{nm}$ will vanish for $m \neq 1$. Therefore, the original boundary conditions are decomposed into $4N$ equations limited only by n. For a given n, we have four linear equations containing the Mie coefficients:

$$\begin{aligned}j_n(z)d_n &= b_n h_n^{(1)}(x) + j_n(x) \\ \mu_0[zj_n(z)]'d_n &= b_n \mu[xh_n^{(1)}(x)]' + \mu[xj_n(x)]' \\ \mu_0 m j_n(z)c_n &= a_n \mu h_n^{(1)}(x) + \mu j_n(x) \\ \frac{1}{m}[zj_n(z)]'c_n &= a_n[xh_n^{(1)}(x)]' + [xj_n(x)]'\end{aligned}, \qquad (2.17)$$

in which the prime indicates differentiation with respect to the argument in parentheses. The size parameter describes the ratio of the sphere radius to the wavelength, $x = 2\pi R/\lambda$ and $z = mx$, where m is the relative refractive index, defined by $m = n'/n_B$ ($n' = \sqrt{\varepsilon/\varepsilon_0}$ is the refractive index of the sphere and n_B is the refractive index of the surrounding medium). Then the Mie coefficients become

$$\begin{aligned}a_n &= \frac{\mu j_n(x)[zj_n(z)]' - \mu_0 m^2 j_n(z)[xj_n(x)]'}{\mu_0 m^2 j_n(z)[xh_n^{(1)}(x)]' - \mu h_n^{(1)}(x)[zj_n(z)]'} \\ b_n &= \frac{\mu j_n(z)[xj_n(x)]' - \mu_0 j_n(x)[zj_n(z)]'}{\mu_0 h_n^{(1)}(x)[zj_n(z)]' - \mu j_n(z)[xh_n^{(1)}(x)]'}\end{aligned}, \qquad (2.18)$$

and

$$c_n = \frac{\mu m j_n(x)[xh_n^{(1)}(x)]' - \mu m h_n^{(1)}(x)[xj_n(x)]'}{\mu_0 m^2 j_n(z)[xh_n^{(1)}(x)]' - \mu h_n^{(1)}(x)[zj_n(z)]'}$$
$$d_n = \frac{\mu h_n^{(1)}(x)[xj_n(x)]' - \mu j_n(x)[xh_n^{(1)}(x)]'}{\mu_0 h_n^{(1)}(x)[zj_n(z)]' - \mu j_n(z)[xh_n^{(1)}(x)]'}$$

(2.19)

Figure 2.2 shows examples of the EM field distribution around a silver sphere in water, calculated from Eq. 2.14, together with the expansion coefficient given by Eqs. 2.9, 2.10, 2.18, and 2.19. The energy flows are also indicated by the black lines. Obviously, the local electric fields near the particle surfaces are enhanced with respect to the incident plane wave. For the smaller NP, the response is nearly pure dipolar so the local field is dipole like. On the contrary, the incident light is strongly diffracted by the larger particle in Fig. 2.2b, creating a clear interference pattern around the particle. For both particles, some of the energy flow is stopped inside the particles, indicating the absorption due to the Ohmic loss in the metal.

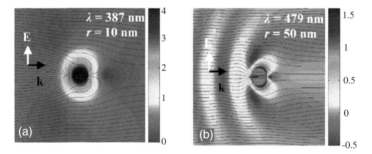

Figure 2.2 Local electric field intensity ($|E|^2/|E_0|^2$) distribution calculated by Eq. 2.14, for silver NPs in water ($n = 1.33$). The incident wave vector and electric field polarization are indicated by the black and white arrows, respectively. The color bar is in logarithmic scale. Reprinted from Ref. [8], Copyright (2004), with permission from Elsevier.

2.1.4 Optical Far-Field Cross Sections

The strengths of the scattering and absorption by a particle are characterized by the so-called scattering/absorption cross sections, which can be qualitatively measured by, for example, dark-field spectroscopy or an absorption spectrometer. In electromagnetism, the far-field cross sections of an object in a homogeneous medium can be calculated via an integral of energy flow over an arbitrary surface σ enclosing the obstacles as

$$\sigma_{scat} = \frac{1}{2I_0}\text{Re}\int_\sigma (\mathbf{E}_s \times \mathbf{H}_s^*) \cdot \hat{\mathbf{n}} d\sigma, \qquad (2.20)$$

and

$$\sigma_{abs} = -\frac{1}{2I_0}\text{Re}\int_\sigma (\mathbf{E}_{total} \times \mathbf{H}_{total}^*) \cdot \hat{\mathbf{n}} d\sigma, \qquad (2.21)$$

where the term $\frac{1}{2}\text{Re}(\mathbf{E}\times\mathbf{H}^*)$ is the Poynting vector and $I_0 = \frac{1}{2}\text{Re}(\sqrt{\varepsilon\mu^{-1}})|\mathbf{E}_0(\mathbf{r})|^2$ is the incident field intensity. According to energy conservation, any energy flowing into a volume should flow out for an incident plane wave, that is, $\text{Re}\int_\sigma (\mathbf{E}_i \times \mathbf{H}_i^*) \cdot \hat{\mathbf{n}} d\sigma = 0$. By definition, the extinction cross section is the sum of the scattering and absorption cross sections and is given by

$$\sigma_{ext} = -\frac{1}{2I_0}\text{Re}\int_\sigma (\mathbf{E}_i \times \mathbf{H}_s^* + \mathbf{E}_s \times \mathbf{H}_i^*) \cdot \hat{\mathbf{n}} d\sigma. \qquad (2.22)$$

It's convenient to carry the integral over a spherical surface with an infinite radius so that the asymptotic behavior of the spherical Bessel functions can be used. For $r \to \infty$, $\frac{\partial[rh_n^{(1)}(kr)]}{\partial r} \approx \frac{e^{ikr}}{i^n}$ and $\frac{\partial[rj_n(kr)]}{\partial r} \approx i\cos\left(kr - \frac{n+1}{2}\pi\right)$. Knowing that the r^{-2} terms vanish compared to the r^{-1} terms and after tedious deduction, one finally obtains the far-field cross sections as

$$\sigma_{scat} = \frac{2\pi}{k^2}\sum_{n=1}^\infty (2n+1)\left(|a_n|^2 + |b_n|^2\right)$$

$$\sigma_{ext} = -\frac{2\pi}{k^2}\sum_{n=1}^\infty (2n+1)\text{Re}(a_n + b_n) \qquad (2.23)$$

The absorption cross section is simply obtained by taking the scattering out of the extinction. Far-field efficiencies Q, defined as the ratio of cross sections to the geometrical cross section of the scatterers, can be used to compare the ability of extinction/

scattering/absorption to the external field for different scatterers. For a single spherical scatterer, the geometrical cross section equals πa^2. Thus, we have

$$Q_{scat} = \frac{2}{x^2}\sum_{n=1}^{\infty}(2n+1)\left(|a_n|^2 + |b_n|^2\right)$$

$$Q_{ext} = -\frac{2}{x^2}\sum_{n=1}^{\infty}(2n+1)\mathrm{Re}(a_n + b_n) \quad . \tag{2.24}$$

$$Q_{abs} = Q_{ext} - Q_{scat}$$

Figure 2.3 shows examples of far-field efficiencies calculated from Eq. 2.24 for the same silver NPs shown in Fig. 2.2. The calculated spectra reveal the surface plasmon resonance properties of the investigated particles, including the peak positions, the magnitudes, and the width of the surface plasmon resonance. When the radius of the NP is increased from 10 nm to 50 nm, the dipolar peak is red-shifted from 387 nm to 479 nm, with its width broadened due to the increased radiative damping. Quadruple resonance, which is absent in small NPs, appears in larger NPs. What's more, the absorption due to Ohmic heating is larger than the scattering for smaller NPs but the opposite is true for larger ones.

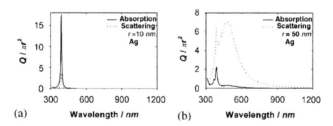

Figure 2.3 Optical far-field cross sections for silver NPs in water, the same ones as in Fig. 2.2. Reprinted from Ref. [8], Copyright (2004), with permission from Elsevier.

2.2 Generalized Mie Theory

Conventional Mie theory was restricted to scattering by one sphere, until Stein (1961) [9] and later Cruzan (1962) [10] presented the addition theorems for VSH. The translation coefficients between two coordinates involved a Wigner 3-*j* symbol, which is computational time consuming. The landmark works by Liang and Lo in 1967 [11], by introducing addition theorems for VSH, extended conventional Mie theory to the two-sphere system in a formal form. Later, Bruning and Lo derived a recursive formula for the translation coefficients in the absence of the Wigner 3-j symbol, which greatly encouraged the computational methods at that time [12]. In 1991, Fuller introduced the order-of-scattering (OS) method to trace the light between two particles [13]. But the OS method becomes computation consuming when resonances occur.

Beside the number of spheres, conventional Mie theory has been extended into different cases, known as generalized Mie theory. Generally, the extension can be grouped into five different aspects: (i) Gaussian incident beams [14, 15], (ii) particles with an inhomogeneous or anisotropic material response [16], (iii) spheroid particles [17, 18], (iv) coated or multilayered spheres [19–22], and (v) multiple spheres, as mentioned before. In the following section, we will introduce the extensions for the last two cases (see Refs. [23–28] and references therein).

2.2.1 Scattering by a Multilayered Sphere

Analogous to the multishell semiconductor quantum dots for confining the motions of electrons, the multilayered particles enable manipulating the optical flow (photons) in a desired manner. Proper combination of material and structural parameters provides a unique way for achieving some interesting optical response. Among those, core-shell NPs receive special attention due to their controlled synthesis [29] and several important applications, such as optical labeling and cancer therapy [30–32]. The EM response of such concentrated sphere can be analytically obtained by solving the boundary value equations over two spherical interfaces and replacing the original scattering coefficients, a_n and b_n, with extended ones, a'_n and b'_n. The solution was first obtained by Aden and Kerker [19] in 1951 and then in 1952 by Güttler [20]. Using a similar method, Bhandari [21] made an extension to a single sphere with an arbitrary number of layers. Recently, Sinzig and Quinten [22] presented a formalism that allows concise recursive calculations of the scattering coefficients for multilayered spheres. Except the complexity of the scattering coefficients for the coated particles, the solving procedures follow those in conventional Mie theory.

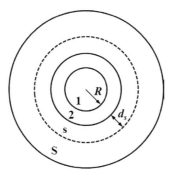

Figure 2.4 Schematic drawing of a coated particle with (S − 1) layers.

Suppose a sphere with a radius R is coated with $(S-1)$ different layers, with thickness d_s, and refractive index n_s for the sth layer. We now have S boundaries, as shown in Fig. 2.4. In the embedding medium and the core, the EM field is expanded by VSH $|nm3p\rangle$ and $|nm1p\rangle$, respectively. In the shell layers, however, the EM field may be represented by both $|nm1p\rangle$ and $|nm2p\rangle$ since both of them are singularity free in these regions. After applying the boundary conditions on the S interfaces, we have $4S$ linear equations similar to Eq. 2.17. According to Sinzig and Quinten [22], the scattering coefficients can be recursively obtained as follows:

$$a_n' = -\frac{m_s\psi_n(m_s y_s)[\psi_n'(y_s)+T_n^S \chi_n'(y_s)]-\psi_n'(m_s y_s)[\psi_n(y_s)+T_n^S \chi_n(y_s)]}{m_s\xi_n(m_s y_s)[\psi_n'(y_s)+T_n^S \chi_n'(y_s)]-\xi_n'(m_s y_s)[\psi_n(y_s)+T_n^S \chi_n(y_s)]}$$

$$b_n' = -\frac{\psi_n(m_s y_s)[\psi_n'(y_s)+D_n^S \chi_n'(y_s)]-m_s\psi_n'(m_s y_s)[\psi_n(y_s)+D_n^S \chi_n(y_s)]}{\xi_n(m_s y_s)[\psi_n'(y_s)+D_n^S \chi_n'(y_s)]-m_s\xi_n'(m_s y_s)[\psi_n(y_s)+D_n^S \chi_n(y_s)]}$$

(2.25)

where $m_s = k_{s+1}/k_s$, $y_s = k_s r_s$ $\left(r_s = R + \sum_{i=2}^{S} d_s\right)$, $\psi_n(x) = xj_n(x)$, $\chi_n(x) = xy_n(x)$, $\xi_n(x) = xh_n^{(1)}(x)$, and $k_s = (2\pi n_s)/\lambda$ and the terms T_n^S and D_n^S are given by

$$T_n^S = -\frac{m_s\psi_n(m_s y_s)[\psi_n'(y_s)+T_n^{S-1} \chi_n'(y_s)]-\psi_n'(m_s y_s)[\psi_n(y_s)+T_n^{S-1} \chi_n(y_s)]}{m_s\chi_n(m_s y_s)[\psi_n'(y_s)+T_n^{S-1} \chi_n'(y_s)]-\chi_n'(m_s y_s)[\psi_n(y_s)+T_n^{S-1} \chi_n(y_s)]}$$

$$D_n^S = -\frac{\psi_n(m_s y_s)[\psi_n'(y_s)+D_n^{S-1} \chi_n'(y_s)]-m_s\psi_n'(m_s y_s)[\psi_n(y_s)+D_n^{S-1} \chi_n(y_s)]}{\chi_n(m_s y_s)[\psi_n'(y_s)+D_n^{S-1} \chi_n'(y_s)]-m_s\chi_n'(m_s y_s)[\psi_n(y_s)+D_n^{S-1} \chi_n(y_s)]}$$

(2.26)

When $s \leq 0$, T_n^s and $D_n^s = 0$. Here, the magnetic permeabilities of the different materials are all assumed to be equal to that of vacuum, μ_0.

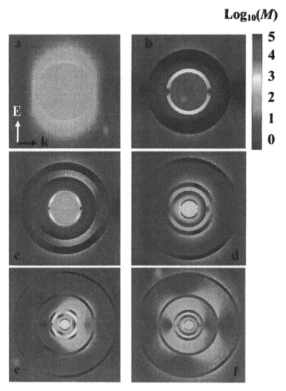

Figure 2.5 EM intensity enhancement ($M = |\mathbf{E}|^2/|\mathbf{E}_0|^2$) distribution around a multilayered nanosphere. The polarization and wave vector of the incident plane wave is indicated by the white and black arrow, respectively. The radius of the silver core is $R = 10$ nm. The thicknesses of the Ag shells are $d_2 = 10$ nm, $d_4 = 5$ nm, $d_6 = 5$ nm, and $d_8 = 5$ nm. The separations between the neighboring silver are $d_1 = 2$ nm, $d_3 = 5$ nm, $d_5 = 25$ nm, and $d_7 = 40$ nm. The incident wavelength λ for (a–f) is 358, 690, 664, 665, 678, and 888 nm, respectively. Reprinted with permission from Ref. [25]. Copyright (2005) by the American Physical Society.

Representative examples of concentric NPs for tunable nanophotonic architectures are shown in Fig. 2.5. A single NP with 0, 2, 4, 6, 8, and 10 layers is compared. The coating layers provide an additional degree for tailoring the local EM field, such as the field enhancement, the distribution, and the spectral response. With the development of modern nanotechnologies, scientists are now able to fabricate multilayer NPs with the desired size and thickness [33] and thus create an additional degree for achieving nanophotonic tailoring.

2.2.2 Addition Theorem for VSH

Extending Mie theory to two or more spheres requires a translation of the VSH between coordinates centered at different spheres, which is accomplished by the so-called addition theorem for VSH. The outward-going electric and magnetic modes, $|nm31,s\rangle$ and $|nm32,s\rangle$, in the sth sphere can be expanded into the modes $|\mu\nu 11,l\rangle$ and $|\mu\nu 12,l\rangle$ centered in the coordinates of the lth sphere [9, 10] as

$$|nm31,s\rangle = \sum_{\nu=1}^{\infty}\sum_{\mu=-\nu}^{\nu}\left(A_{\mu\nu}^{mn\ l}|\mu\nu 11,l\rangle + B_{\mu\nu}^{mn\ l}|\mu\nu 12,l\rangle\right)$$

$$|nm32,s\rangle = \sum_{\nu=1}^{\infty}\sum_{\mu=-\nu}^{\nu}\left(B_{\mu\nu}^{mn\ l}|\mu\nu 11,l\rangle + A_{\mu\nu}^{mn\ l}|\mu\nu 12,l\rangle\right)$$

(2.27)

The translation coefficients $A_{\mu\nu}^{mn}$ and $B_{\mu\nu}^{mn}$ are given by

$$A_{\mu\nu}^{mn} = (-1)^{\mu} i^{\nu-n} \frac{2\nu+1}{2\nu(\nu+1)} \sum_{p=|n-\nu|}^{n+\nu} (-i)^{p} [n(n+1)+\nu(\nu+1)-p(p+1)]$$

$$\times a(m,n,-\mu,\nu,p)\xi_{p}(kd_{s,l})P_{p}^{m-\mu}(\cos\theta_{s,l})\exp[i(m-\mu)\phi_{s,l}]$$

$$B_{\mu\nu}^{mn} = (-1)^{\mu} i^{\nu-n} \frac{2\nu+1}{2\nu(\nu+1)} \sum_{p=|n-\nu|}^{n+\nu} (-i)^{p} b(m,n,-\mu,\nu,p,p-1)\xi_{p}(kd_{s,l})$$

$$P_{p}^{m-\mu}(\cos\theta_{s,l})\exp[i(m-\mu)\phi_{s,l}]$$

(2.28)

where $d_{s,l}$ is the line segment joining the centers of sphere s and l, $\theta_{s,l}$ is the angle between $d_{s,l}$ and the z axis of sphere s, $\phi_{s,l}$ corresponds to the azimuthal coordinate of the lth sphere in the sth coordinate system, and $\xi_p(x) = x h_p^{(1)}(x)$. The $b(m, n, -\mu, \nu, p, p-1)$ terms are defined as

$$b(m,n,-\mu,\nu,p,p-1) = \frac{2p+1}{2p-1}\{(\nu-\mu)(\nu+\mu+1)a(m,n,-\mu-1,\nu,p-1)$$
$$-(p-m+\mu)(p-m-\mu+1)a(m,n,-\mu+1,\nu,p-1)$$
$$+2\mu(p-m+\mu)a(m,n,-\mu,\nu,p-1)\}$$

.

The $a(m, n, \mu, v, p)$ terms are defined by

$$P_n^m(\cos\theta)P_v^\mu(\cos\theta) = \sum_{p=|n-v|}^{n+v} a(m,n,\mu,v,p)P_p^{m+\mu}(\cos\theta),$$

where

$$a(m,n,\mu,v,p) = (2p+1)\frac{(p-m-\mu)!}{(p+m+\mu)!}\int_{-1}^{1} P_n^m(x)P_v^\mu(x)P_p^{m+\mu}(x)dx$$

$$= (-1)^{m+\mu}(2p+1)\left[\frac{(n+m)!(v+\mu)!(p-m-\mu)!}{(n-m)!(v-\mu)!(p+m+\mu)!}\right]^{1/2}$$

$$\begin{bmatrix} n & v & p \\ 0 & 0 & 0 \end{bmatrix}\begin{bmatrix} n & v & p \\ m & \mu & -m-\mu \end{bmatrix}$$

and $\begin{bmatrix} j_1 & j_2 & j_3 \\ m_1 & m_2 & m_3 \end{bmatrix}$ is the Wigner 3-j symbol, which is very time consuming to calculate. However, when the two coordinate systems are chosen so that both $\theta_{s,l}$ and $\varphi_{s,l}$ are equal to 0, the translation coefficients $A_{\mu v}^{mn}$ and $B_{\mu v}^{mn}$ are equal to 0 if $m \neq \mu$. When the two coordinates are chosen as $\theta_{s,l} = \pi$ (reversed translation) and $\varphi_{s,l} = 0$, the translation coefficients are still equal to 0 if $m \neq \mu$ but A_{mv}^{mn} and B_{mv}^{mn} are preceded by factors $(-1)^{n+v}$ and $(-1)^{n+v+1}$, respectively. These translation coefficients can be simplified as [12]

$$A_{mv}^{mn} = (-1)^m i^{v-n}\frac{2v+1}{2v(v+1)}\sum_{p=|n-v|}^{n+v}(-i)^p[n(n+1)+v(v+1)-p(p+1)]$$

$$\times a(m,n,-m,v,p)\begin{bmatrix} h_p^{(1)}(kd_{s,l}) \\ j_p(kd_{s,l}) \end{bmatrix}$$

(2.29)

$$B_{mv}^{mn} = (-1)^m i^{v-n}\frac{2v+1}{2v(v+1)}\sum_{p=|n-v|}^{n+v}(-i)^p(-2imkd_{s,l})$$

$$a(m,n,-m,v,p)\begin{bmatrix} h_p^{(1)}(kd_{s,l}) \\ j_p(kd_{s,l}) \end{bmatrix}, \text{ for } \begin{bmatrix} a \leq d_{s,l} \\ a > d_{s,l} \end{bmatrix},$$

where the coefficient $a(m, n, -m, v, p)$, denoted as a_p, has a relatively simple three-term recurrence relation, as shown by Bruning and Lo [12], which allows for an expedient calculation as follows:

$$\alpha_{p-3}a_{p-4} - (\alpha_{p-2} + \alpha_{p-1} - 4m^2)a_{p-2} + \alpha_p a_p = 0, \qquad (2.30)$$

where

$$p = n+v, n+v-2, \cdots, |n-v|, \quad \alpha_p = \frac{\left[(n+v+1)^2 - p^2\right]\left[p^2 - (n-v)^2\right]}{4p^2 - 1},$$

and

$$a_{n+v} = \frac{(2n-1)!!(2v-1)!!}{(2n+2v-1)!!} \frac{(n+v)!}{(n-m)!/(v+m)!}$$

$$a_{n+v-2} = \frac{(2n+2v-3)}{(2n-1)(2v-1)(n+v)}[nv - m^2(2n+2v-1)]a_{n+v}$$

2.2.3 Order-of-Scattering Method for Two Spheres

Scattering light coming from spheres A and B is marked by *a* and *b* for convenience. As the schematic shows in Fig. 2.6, scattering light from each sphere can be divided into different scattering orders. The first order of the scattered field by sphere A contains two parts: one is the scattering light of the incident planar wave directly by sphere A, represented by *a*, and the other is from the incident light scattered by sphere B and then by sphere A, which is now denoted by *bTa*. *T* is the translation matrix between the coordinate centered at B to that centered at A. Therefore, the first OS from sphere A can be represented by *a* + *bTa*.

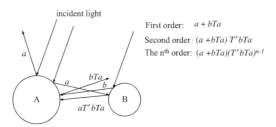

Figure 2.6 Schematic drawing of the scattering orders for two particles.

The first OS coming from sphere A, when scattered by sphere B and then again by sphere A, forms the second OS light, marked as (*a* + *bTa*)*T'bTa*. Analogically, the *n*th OS forms when the first OS is reflected first by sphere A and then by sphere B for *n* − 1 times.

Hence, it could be signed by $(a + bTa)(T'bTa)^{n-1}$. The scattering field of sphere A is the summation of different scattering orders. Using the summation formula of geometric sequence, the scattering field is given by

$$E_A^s = \sum_{n=1}^{\infty} (a+bTa)(T'bTa)^{n-1} = \frac{a+bTa}{1-T'bTa}. \quad (2.31)$$

In analogy to the above equation, the scattering field outward-going from sphere B is given as follows:

$$E_B^s = \frac{b+aT'b}{1-TaT'b}. \quad (2.32)$$

The total scattered electric field of the dimer is, however, given by $E^s = E_A^s + E_B^s$. To facilitate the evaluation of the scattered field, Eqs. 2.31 and 2.32 should be matriculated in a manner that can be easily implemented by a computer. A schedule of matriculation of the above OS method for a dimer system would be presented in the next section, together with those for the multiple-sphere system.

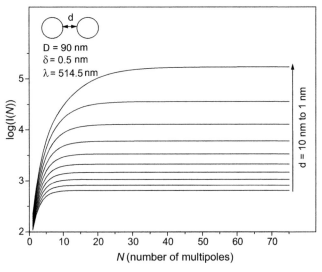

Figure 2.7 Calculated intensity I as a function of the number of multipoles N for a point at the dimer axis between two Ag spheres δ = 0.5 nm from one of the spherical surfaces. The diameters of the spheres are 90 nm, and the polarization of light is parallel to the dimer's axis. The calculations are performed for different separation distances d, from 1 nm to 10 nm. Reproduced from Ref. [34], with permission from SPIE.

For practical calculations, one needs to use a suitable number of multipoles $n \in [1, N]$ to balance the accuracy and the calculation complexity. When the size of the spheres increases, or the distance between them decreases, a large N should be used [35]. As an example of how to choose a reasonable number of multipole terms, Fig. 2.7 shows the intensity I versus the number of multipoles N used in a calculation for a dimer with different separation distances. For a separation distance of around 10 nm between larger particles (90 nm in diameter), it is enough to include n up to 10 to get convergence, whereas for distances down to 1 nm, it is necessary to include more than 50 multipole terms.

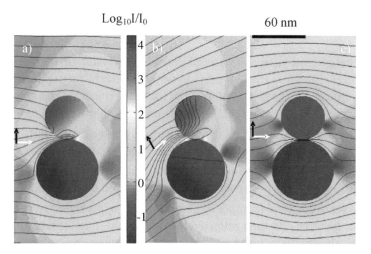

Figure 2.8 Local intensity distribution I/I_0 in the logarithmic scale and the streams of the EM energy (solid lines) in the plane of the wave vector **k** (the white arrow) and the electric field **E** (the black arrow) through the centers of a Au sphere (upper) with a radius of 25 nm and a Ag sphere (lower) with a radius of 35 nm at the incident wavelength of 514.5 nm for (a) and (b) and 800 nm for (c). **k** is perpendicular to the axis of the two spheres in (a) and (c). In (b), **k** and the axis of the two spheres have an angle of 60°. The surface separation between the two spheres is 1 nm. Less than this distance, Mie theory will probably fail due to the spurted electrons to each other. The dielectric function of Au and Ag is obtained from Johnson and Christy (JC) [38]. Reprinted from Ref. [26], Copyright (2003), with permission from Elsevier.

For two closely separated spheres, the scattered light from one sphere is reflected back and forth by the other sphere. Such multiple scattering processes dramatically raise the EM field in the gap when resonance happens. A typical OS method may run into a convergence problem at this case. The above method, however, avoids this problem by replacing the sum over different scattering orders by matrix inversion (see Eq. 2.31). Figure 2.8 shows a Au-Ag nanosphere heterodimer excited by a plane wave with an electric field indicated by the arrows. Apparently, the EM field is concentrated into the interparticle gap, which contributes to various nonlinear optical effects, such as surface-enhanced Raman scattering (SERS) [35, 36] and high harmonics generations [37].

2.2.4 Order-of-Scattering Method for an Arbitrary Number of Spheres

In the general case, we need to consider a cluster of L different spheres. Similar to the two-particle system, one arbitrary sphere, for example, the lth sphere, is picked and the scattering from this sphere is denoted as a_l. The total scattering from the other $L - 1$ spheres is denoted as b_l^{L-1} (those $L - 1$ spheres are thus treated as one unit). The first order of the scattering from the lth sphere is composed of a_l (the scattered fields from the incident light directly) and $b_l^{L-1} T_l a_l$ (the light scattered by the other spheres and then scattered by the lth sphere). This scattered field (composed of a_l and $b_l^{L-1} T_l a_l$) is reflected by the other spheres and then reflected by the lth sphere to infinite order. Thus, the total scattered field from the lth sphere can be denoted as $\dfrac{a_l + b_l^{L-1} T_l a_l}{1 - b_l^{L-1} T_l a_l T_l'}$, where $(1 - b_l^{L-1} T_l a_l T_l')^{-1} = \sum\limits_{n=1}^{\infty} (b_l^{L-1} T_l a_l T_l')^{n-1}$. The total scattered field is the sum of the scattered fields from the lth sphere and the other $L - 1$ spheres:

$$E^s = \frac{a_l + b_l^{L-1} T_l a_l}{1 - b_l^{L-1} T_l a_l T_l'} + \frac{b_l^{L-1} + a_l T_l' b_l^{L-1}}{1 - a_l T_l' b_l^{L-1} T_l}. \tag{2.33}$$

In a similar way, if we choose one sphere, for example, the qth sphere within the $L - 1$ cluster, the b_l^{L-1} can then be described as

$$b_l^{L-1} = \frac{a_q + b_q^{L-2} T_q a_q}{1 - b_q^{L-2} T_q a_q T_q'} + \frac{b_q^{L-2} + a_q T_q' b_q^{L-2}}{1 - a_q T_q' b_q^{L-2} T_q}, \text{ regardless of the } l\text{th sphere.}$$

Here, b_q^{L-2} is the scattering from the remaining $L - 2$ spheres except the *l*th and the *q*th spheres. Continue this process for $L - 3$ spheres, $L - 4$ spheres, and so on, until only the last sphere, say, the *p*th sphere, is left. We finally have $b_p^{L-(L-1)} = a_p$. Hence, the total scattering from a multisphere system can be clearly and thoroughly described using the same mathematical formulation as for the two-sphere system. The incident and scattered electric fields for an *L*-sphere system can be written as

$$^i\mathbf{E}_l = \sum_{n=1}^{\infty}\sum_{m=-n}^{n}\sum_{p=1}^{2} {}^i C_{mnp}^l |mn1p\rangle$$

$$^s\mathbf{E} = \bigoplus_{l=1}^{L}{}^s\mathbf{E}_l = \bigoplus_{l=1}^{L}\sum_{n=1}^{\infty}\sum_{m=-n}^{n}\sum_{p=1}^{2} {}^s C_{mnp}^l |mn3p\rangle$$

(2.34)

where the symbol $\oplus$ means that the sum is carried in Cartesian coordinates. According to the above-mentioned recursive OS method, the expansion coefficients ${}^s C_{mnp}^l$ can be constructed from the incidence coefficients ${}^i C_{mnp}^l$, the Mie coefficients a_n^l and b_n^l, and the translation coefficients ${}^{lh}A_{\mu\nu}^{mn}$ and ${}^{lh}B_{\mu\nu}^{mn}$ between the sphere *l* and the sphere *h*. To calculate the incident/scattered EM field, a matriculation process to Eq. 2.34 will be introduced in the following section.

First of all, let's define a matrix X_p^l ($p = 1, 2$) containing the incidence coefficients C_{mnp}^l, a matrix Y_{jp}^l from the VSH $|mnjp\rangle^l \equiv |mnjp,l\rangle$, and matrices Z_i^l and U_i^l ($i = 1, 2$) from the Mie coefficients, centered in the *l*th sphere, as

$$X_p^l = [C_{-11p}^l \ C_{01p}^l \ C_{11p}^l \ C_{-22p}^l \ C_{-12p}^l \ C_{02p}^l$$

$$C_{12p}^l \ C_{22p}^l \cdots C_{-NNp}^l \ C_{-(N-1)Np}^l \cdots C_{NNp}^l]$$

$$Y_{jp}^l = [|-11jp\rangle^l \ |01jp\rangle^l \ |11jp\rangle^l \ |-22jp\rangle^l ,$$

$$\cdots |22jp\rangle^l \cdots |-NNjp\rangle^l \cdots |NNjp\rangle^l]$$

$$Z_1^l = [b_1^l \ b_1^l \ b_1^l \ b_2^l \ b_2^l \ b_2^l \ b_2^l \ b_2^l \cdots b_N^l \cdots b_N^l]^D ,$$

$$Z_2^l = [a_1^l \ a_1^l \ a_1^l \ a_2^l \ a_2^l \ a_2^l \ a_2^l \ a_2^l \cdots a_N^l \cdots a_N^l]^D ,$$

$$U_1^l = [d_1^l \ d_1^l \ d_1^l \ d_2^l \ d_2^l \ d_2^l \ d_2^l \ d_2^l \cdots d_N^l \cdots d_N^l]^D ,$$

and

$$U_2^l = [c_1^l \ c_1^l \ c_1^l \ c_2^l \ c_2^l \ c_2^l \ c_2^l \ c_2^l \cdots c_N^l \cdots c_N^l]^D.$$

The superscript D transfers the elements in the vector into the diagonal elements in the matrix with zero off-diagonal elements. The translation coefficients from the sth sphere to the lth sphere are matriculated into two $(N^2 + 2N) \times (N^2 + 2N)$ matrices:

$$^{ls}\overline{A} = \begin{bmatrix} ^{ls}A_{-11}^{-11} & ^{ls}A_{01}^{-11} & ^{ls}A_{11}^{-11} & ^{ls}A_{-22}^{-11} & \cdots & ^{ls}A_{22}^{-11} & \cdots & ^{ls}A_{-NN}^{-11} & \cdots & ^{ls}A_{NN}^{-11} \\ ^{ls}A_{-11}^{01} & ^{ls}A_{01}^{01} & ^{ls}A_{11}^{01} & ^{ls}A_{-22}^{01} & \cdots & ^{ls}A_{22}^{01} & \cdots & ^{ls}A_{-NN}^{01} & \cdots & ^{ls}A_{NN}^{01} \\ ^{ls}A_{-11}^{11} & ^{ls}A_{01}^{11} & ^{ls}A_{11}^{11} & ^{ls}A_{-22}^{11} & \cdots & ^{ls}A_{22}^{11} & \cdot & ^{ls}A_{-NN}^{11} & \cdot & ^{ls}A_{NN}^{11} \\ ^{ls}A_{-11}^{-22} & ^{ls}A_{01}^{-22} & ^{ls}A_{11}^{-22} & ^{ls}A_{-22}^{-22} & \cdots & ^{ls}A_{22}^{-22} & \cdots & ^{ls}A_{-NN}^{-22} & \cdots & ^{ls}A_{NN}^{-22} \\ \vdots & \cdot & \cdot & \vdots & \cdot & \vdots & \cdot & \vdots & \cdot & \vdots \\ ^{ls}A_{-11}^{22} & \cdots & \cdots & ^{ls}A_{-22}^{22} & \cdots & ^{ls}A_{22}^{22} & \cdots & ^{ls}A_{-NN}^{22} & \cdots & ^{ls}A_{NN}^{22} \\ \vdots & \cdot & \cdot & \vdots & \cdot & \vdots & \cdot & \vdots & \cdot & \vdots \\ ^{ls}A_{-11}^{-NN} & \cdots & \cdots & ^{ls}A_{-22}^{-NN} & \cdots & ^{ls}A_{22}^{-NN} & \cdots & ^{ls}A_{-NN}^{-NN} & \cdots & ^{ls}A_{NN}^{-NN} \\ \vdots & \cdot & \cdot & \vdots & \cdot & \vdots & \cdot & \vdots & \cdot & \vdots \\ ^{ls}A_{-11}^{NN} & \cdots & \cdots & ^{ls}A_{-22}^{NN} & \cdots & ^{ls}A_{22}^{NN} & \cdots & ^{ls}A_{-NN}^{NN} & \cdots & ^{ls}A_{NN}^{NN} \end{bmatrix}$$

and

$$^{ls}\overline{B} = \begin{bmatrix} ^{ls}B_{-11}^{-11} & ^{ls}B_{01}^{-11} & ^{ls}B_{11}^{-11} & ^{ls}B_{-22}^{-11} & \cdots & ^{ls}B_{22}^{-11} & \cdots & ^{ls}B_{-NN}^{-11} & \cdots & ^{ls}B_{NN}^{-11} \\ ^{ls}B_{-11}^{01} & ^{ls}B_{01}^{01} & ^{ls}B_{11}^{01} & ^{ls}B_{-22}^{01} & \cdots & ^{ls}B_{22}^{01} & \cdots & ^{ls}B_{-NN}^{01} & \cdots & ^{ls}B_{NN}^{01} \\ ^{ls}B_{-11}^{11} & ^{ls}B_{01}^{11} & ^{ls}B_{11}^{11} & ^{ls}B_{-22}^{11} & \cdots & ^{ls}B_{22}^{11} & \cdots & ^{ls}B_{-NN}^{11} & \cdot & ^{ls}B_{NN}^{11} \\ ^{ls}B_{-11}^{-22} & ^{ls}B_{01}^{-22} & ^{ls}B_{11}^{-22} & ^{ls}B_{-22}^{-22} & \cdots & ^{ls}B_{22}^{-22} & \cdots & ^{ls}B_{-NN}^{-22} & \cdots & ^{ls}B_{NN}^{-22} \\ \vdots & \cdot & \cdot & \vdots & \cdot & \vdots & \cdot & \vdots & \cdot & \vdots \\ ^{ls}B_{-11}^{22} & \cdots & \cdots & ^{ls}B_{-22}^{22} & \cdots & ^{ls}B_{22}^{22} & \cdots & ^{ls}B_{-NN}^{22} & \cdots & ^{ls}B_{NN}^{22} \\ \vdots & \cdot & \cdot & \vdots & \cdot & \vdots & \cdot & \vdots & \cdot & \vdots \\ ^{ls}B_{-11}^{-NN} & \cdots & \cdots & ^{ls}B_{-22}^{-NN} & \cdots & ^{ls}B_{22}^{-NN} & \cdots & ^{ls}B_{-NN}^{-NN} & \cdots & ^{ls}B_{NN}^{-NN} \\ \vdots & \cdot & \cdot & \vdots & \cdot & \vdots & \cdot & \vdots & \cdot & \vdots \\ ^{ls}B_{-11}^{NN} & \cdots & \cdots & ^{ls}B_{-22}^{NN} & \cdots & ^{ls}B_{22}^{NN} & \cdots & ^{ls}B_{-NN}^{NN} & \cdots & ^{ls}B_{NN}^{NN} \end{bmatrix}.$$

Then, we are able to construct the following matrices:

$$G_l = [X_1^l \ X_2^l],$$

$$^l W_1^E = [Y_{11}^l \ Y_{12}^l]^T, \ ^l W_3^E = [Y_{31}^l \ Y_{32}^l]^T, \ ^l W_3^H = [Y_{32}^l \ Y_{31}^l]^T,$$

$$S_l = \begin{bmatrix} Z_1^l & 0 \\ 0 & Z_2^l \end{bmatrix}, \quad P_l = \begin{bmatrix} U_1^l & 0 \\ 0 & U_2^l \end{bmatrix},$$

and

$$\Omega_{ls} = \begin{pmatrix} {}^{ls}\overline{A} & {}^{ls}\overline{B} \\ {}^{ls}\overline{B} & {}^{ls}\overline{A} \end{pmatrix}.$$

The superscript T means a transposed matrix. For an arbitrary-number sphere aggregate, the main task of the calculation is to construct a scattering matrix ${}^L T_l$ for the lth sphere in an L-sphere system. Then, the total scattered field can be given by summing up each scattered component from the individual sphere as

$$\begin{aligned} \mathbf{E}_s &= \sum_{l=1}^{\oplus L} \mathbf{E}_s^l = \sum_{l=1}^{\oplus L} {}^L T_l \cdot {}^l W_3^E \\ \mathbf{H}_s &= \sum_{l=1}^{\oplus L} \mathbf{H}_s^l = \frac{k}{i\omega\mu} \sum_{l=1}^{\oplus L} {}^L T_l \cdot {}^l W_3^H \end{aligned} \quad (2.35)$$

For a single sphere, the scattering matrix can be written as

$$^1 T_1 = G_1 S_1. \quad (2.36)$$

For a two-sphere system, according to the OS approach described earlier, the scattering matrix for the first and second spheres can be given, from Eqs. 2.31 and 2.32, by

$$\begin{aligned} ^2 T_1 &= (G_1 S_1 + G_2 S_2 \Omega_{21} S_1) \sum_{i=0}^{\infty} (\Omega_{12} S_2 \Omega_{21} S_1)^i \\ ^2 T_2 &= (G_2 S_2 + G_1 S_1 \Omega_{12} S_2) \sum_{i=0}^{\infty} (\Omega_{21} S_1 \Omega_{12} S_2)^i \end{aligned} \quad (2.37)$$

This equality can be further formulated into a more compact form:

$$(^2 T_1, {}^2 T_2) = (G_1, G_2) \Psi^{(2)}, \quad (2.38)$$

where

$$\Psi^{(2)} = \begin{pmatrix} S_1 \sum_{i=0}^{\infty} (\Omega_{12} S_2 \Omega_{21} S_1)^i & S_1 \Omega_{12} S_2 \sum_{i=0}^{\infty} (\Omega_{21} S_1 \Omega_{12} S_2)^i \\ S_2 \Omega_{21} S_1 \sum_{i=0}^{\infty} (\Omega_{12} S_2 \Omega_{21} S_1)^i & S_2 \sum_{i=0}^{\infty} (\Omega_{21} S_1 \Omega_{12} S_2)^i \end{pmatrix},$$

and the sum over the infinite series will be replaced by matrix inversion, that is,

$$\sum_{i=0}^{\infty}(\Omega_{kj}S_j\Omega_{jk}S_k)^i = \frac{1}{1-\Omega_{kj}S_j\Omega_{jk}S_k}.$$

For a trimer case, according to the recursive OS described before, the third sphere is irradiated by both the incident light and the scattered light from the first and second spheres. The scattering matrix for the third sphere can be written as

$$^3T_3 = [G_3 S_3 + (G_1, G_2)\Psi^{(2)}\Omega'^{(2)} S_3]\sum_{i=0}^{\infty}(\Omega^{(2)}\Psi^{(2)}\Omega'^{(2)} S_3)^i, \quad (2.39)$$

where $\Omega^{(2)} = (\Omega_{3,1}, \Omega_{3,2})$ and $\Omega'^{(2)} = \begin{pmatrix} \Omega_{1,3} \\ \Omega_{2,3} \end{pmatrix}$.

Then, the scattering matrices for the first and second spheres in the trimer system are similar to the two-sphere system, except a term containing the scattered field from the third sphere:

$$(^3T_1, {}^3T_2) = [(G_1, G_2) + {}^3T_3\Omega^{(2)}]\Psi^{(2)}. \quad (2.40)$$

Equations 2.39 and 2.40 can also be rewritten in a compact form by

$$(^3T_1, {}^3T_2, {}^3T_3) = (G_1, G_2, G_3)\begin{pmatrix} \Psi^{(3)}_{11} & \Psi^{(3)}_{12} & \Psi^{(3)}_{13} \\ \Psi^{(3)}_{21} & \Psi^{(3)}_{22} & \Psi^{(3)}_{23} \\ \Psi^{(3)}_{31} & \Psi^{(3)}_{32} & \Psi^{(3)}_{33} \end{pmatrix}, \quad (2.41)$$

where

$$\Psi^{(3)}_{33} = S_3\sum_{i=0}^{\infty}(\Omega^{(2)}\Psi^{(2)}\Omega'^{(2)} S_3)^i = \frac{S_3}{1-\Omega^{(2)}\Psi^{(2)}\Omega'^{(2)} S_3},$$

$$\Psi^{(3)}_{i3} = \sum_{j=1}^{2}\Psi^{(2)}_{ij}\Omega_{j,3}\Psi^{(3)}_{33}, i = 1,2,$$

$$\Psi^{(3)}_{ij} = \Psi^{(2)}_{ij} + \Psi^{(3)}_{i3}\sum_{k=1}^{2}\Omega_{3,k}\Psi^{(2)}_{kj}, i,j = 1,2,$$

and $\Psi^{(3)}_{3i} = \Psi^{(3)}_{33}\sum_{k=1}^{2}\Omega_{3,k}\Psi^{(2)}_{ki}, i = 1,2$.

Similarly, the scattering matrices for an *L*-sphere system can be deduced as

$$({}^L T_1, {}^L T_2, {}^L T_3, \cdots {}^L T_L) = (G_1, G_2, G_3, \cdots G_L)\Psi^{(L)}, \quad (2.42)$$

where the elements of $\Psi^{(L)}$ are given by

$$\Psi^{(L)}_{LL} = \frac{S_L}{1 - \Omega^{(L-1)}\Psi^{(L-1)}\Omega'^{(L-1)}S_L},$$

$$\Psi^{(L)}_{pL} = \sum_{j=1}^{L-1} \Psi^{(L-1)}_{pj} \Omega_{j,L} \Psi^{(L)}_{LL}, \quad p = 1, 2, \cdots, L-1,$$

$$\Psi^{(L)}_{Lq} = \Psi^{(L)}_{LL} \sum_{j=1}^{L-1} \Omega_{L,j} \Psi^{(L-1)}_{jq}, \quad q = 1, 2, \cdots, L-1,$$

$$\Psi^{(L)}_{pq} = \Psi^{(L-1)}_{pq} + \Psi^{(L)}_{pL} \sum_{j=1}^{L-1} \Omega_{L,j} \Psi^{(L-1)}_{jq}, \quad p, q = 1, 2, \cdots, L-1,$$

$$\Omega^{(L-1)} = [\Omega_{L1}, \Omega_{L2}, \cdots, \Omega_{L,L-1}], \text{ and}$$

$$\Omega'^{(L-1)} = [\Omega_{1L}, \Omega_{2L}, \cdots, \Omega_{L-1,L}]^T.$$

Figure 2.9 gives an example of the local electric field calculation based on the above-mentioned approach for three spheres excited by a linearly polarized plane wave from different directions. When the electric field is parallel to the axis of any of two spheres, the local field intensity rises in the corresponding interparticle gap. No field enhancement is present in the gap when the electric field is perpendicular to the axis. Similar to the two-sphere system, the giant field enhancement at the junctions in such particle aggregate creates multiple hot spots, enabling weak optical signal detections such as single-molecule SERS [39].

Besides tuning the field enhancement, a multiple-particle system can also be designed to rotate the polarization of the Raman signal of a local probe molecule [40], to create cascaded field enhancement in a similar chain [41], and to enable subwavelength energy transport [42], etc. Also, by properly designing the number, size, and arrangement of the nanospheres, Fano interference between the dark and bright plasmon resonances would appear, leading to peculiar spectral responses and near-field properties [43, 44].

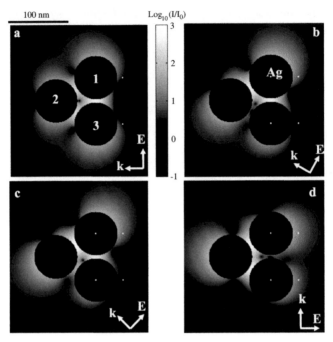

Figure 2.9 Local intensity enhancement distribution I/I_0 in the logarithmic scale in the plane of the wave vector k and the electric field **E**. The Ag spheres are identical to each other, with radius R = 35 nm. The incident wavelength is 514.5 nm, with polarizations illustrated by the arrows of **E**. The dielectric function of Ag is obtained from JC [38]. The number of the multipoles is L = 16. Reprinted from Ref. [27], with permission from The Optical Society.

2.3 Light Scattering by Arbitrarily Shaped Particles and Numerical Simulations

The analytical characteristic of Mie theory ensures that all the EM parameters can be exactly deduced. The only aspect that may limit the precision of the evaluation is the maximum of the multipole included in the calculations. However, Mie theory is restricted to spherical (cylindrical) particles. For an arbitrary-shaped particle, other computational techniques are required. Discrete dipole approximation (DDA), the Green dyadic method (GDM), the T-matrix method, the multiple-multipole method, and the boundary element method are those extensively used in the literature (see Refs. [45–

49] for a review). On the other hand, pure numerical methods, such as the finite-difference time-domain (FDTD) method and the finite element method (FEM), have also been well developed as commercial packages, with a broad range of applications even for complicated systems. In this section, the GDM and pure numerical techniques are briefly introduced in Sections 2.3.1 and 2.3.2, respectively. Comparisons with Mie theory are also concluded.

2.3.1 The Green Dyadic Method

The GDM is also called the (dyadic) Green function method or the Green tensor technique, particularly suitable for solving the diffraction of EM waves by arbitrarily shaped objects embedded in a homogeneous background or near infinitely extended interfaces [50, 51]. The merit of the GDM lies in the fact that only the scatterers have to be discretized, while the background is taken into account by the dyadic Green tensor. For systems in a homogeneous background, the GDM is shown to be equivalent to DDA widely used for investigating arbitrarily shaped NPs [52]. It starts from the Lippmann–Schwinger equation [50]

$$\mathbf{E}(\mathbf{r}) = \mathbf{E}_{inc}(\mathbf{r}) + \int_V d\mathbf{r}' \ddot{G}^B(\mathbf{r},\mathbf{r}') \cdot k_0^2 \Delta\varepsilon(\mathbf{r}')\mathbf{E}(\mathbf{r}'), \qquad (2.43)$$

where $\ddot{G}^B(\mathbf{r},\mathbf{r}')$ is the dyadic Green tensor of the background and $\mathbf{E}_{inc}(\mathbf{r})$ is the incident field. k_0 is the wave number in vacuum, and $\Delta\varepsilon(\mathbf{r}') = \varepsilon(\mathbf{r}') - \varepsilon_B$ is the difference of the dielectric constant inside, $\varepsilon(\mathbf{r}')$, and outside, ε_B, the scatterer. For a homogeneous background, the dyadic Green tensor is given by

$$\ddot{G}^B(\mathbf{r},\mathbf{r}') = \left(1 + \frac{\nabla\nabla}{k_0^2}\right) g(\mathbf{r},\mathbf{r}'), \qquad (2.44)$$

where $g(\mathbf{r},\mathbf{r}') = \dfrac{e^{ik_0|\mathbf{r}-\mathbf{r}'|}}{4\pi|\mathbf{r}-\mathbf{r}'|}$ is the scalar Green function in the free space. The integral over the entire scatterer volume V is taken numerically by discretizing the particle into N subvolume V_j located at $\mathbf{r}_j$. Then, the electric field at the ith subvolume can be rewritten as

$$\mathbf{E}_i = \mathbf{E}_i^{inc} + k_0^2 \sum_{j=1, j\neq i}^{N} \Delta\varepsilon_j \ddot{G}_{ij}^B \cdot \mathbf{E}_j V_j + k_0^2 \Delta\varepsilon_i \vec{\mathbf{M}}_i \cdot \mathbf{E}_i - \frac{\Delta\varepsilon_i}{\varepsilon_B} \ddot{\mathbf{L}} \cdot \mathbf{E}_i, \quad (2.45)$$

where $\vec{\vec{M}}_i = \lim\limits_{\delta V \to 0} \int_{V_i - \delta V} dr' \vec{\vec{G}}^B(\mathbf{r}_i, \mathbf{r}')$ and $\vec{\vec{L}} = \varepsilon_B k_0^2 \lim\limits_{\delta V \to 0} \int_{\delta V} dr' \vec{\vec{G}}^B(\mathbf{r}_i, \mathbf{r}')$.

If the scatterer is discretized into a subvolume centered on a cubic lattice, Eq. 2.45 can be further simplified into a more compacted form as [50]

$$\mathbf{E}_i^{inc} = \sum_j \vec{\vec{A}}_{ij} \mathbf{E}_j, \quad \vec{\vec{A}}_{ij} = \begin{cases} 1 + \dfrac{1}{3}\dfrac{\Delta\varepsilon_j}{\varepsilon_B} - k_0^2 \Delta\varepsilon_j \vec{\vec{M}}, \ i = j \\ -k_0^2 \Delta\varepsilon_j \Delta V \vec{\vec{G}}_{ij}^B, \ i \neq j \end{cases}. \quad (2.46)$$

For $i = 1, \ldots, N$, Eq. 2.46 represents a large system of $3N$ algebraic equations with the $3N$ electric components inside the scatterer as unknowns. After solving Eq. 2.46 by, for example, iterative methods, the field outside the scatterer can be obtained from Eq. 2.45. The far-field cross sections can then be computed using the following formulas:

$$\sigma_{ext} = \dfrac{k_0 \cdot \Delta V}{n_B} \sum_{i=1}^{N} \text{Im}[\Delta\varepsilon(\mathbf{r}_i) \mathbf{E}_i^{inc*} \cdot \mathbf{E}_i]$$

$$\sigma_{abs} = \dfrac{k_0 \cdot \Delta V}{n_B} \sum_{i=1}^{N} \text{Im}(\Delta\varepsilon(\mathbf{r}_i)) \cdot |\mathbf{E}_i|^2 \quad (2.47)$$

$$\sigma_{scat} = \sigma_{ext} - \sigma_{abs}$$

The incident light may be unpolarized, and the particle could be randomly oriented. An average over-the-incident direction and polarization is required. The average extinction (so as scattering and absorption) cross section is given by the following formula:

$$\bar{\sigma}_{ext} = \dfrac{1}{4\pi} \int_0^{2\pi} \int_0^{\pi} \sigma_{ext}(\theta) \sin\theta d\theta d\varphi, \quad (2.48)$$

where $\sigma_{ext}(\theta)$ is the average extinction cross section for s- and p-polarizations and θ is the angle between incident direction and particle axis. An example of the GDM-calculated average extinction spectrum is shown in Fig. 2.10 for a gold nanorod immersed in water. The rod was discretized into $N = 22,036$ subvolumes sitting on a grid with the lattice constant as small as 0.5 nm, as shown in the inset.

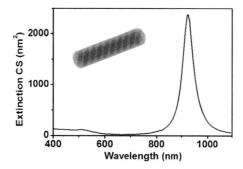

Figure 2.10 Extinction cross section of a gold nanorod in water. The rod is represented by N = 22,036 grids in the calculation, as shown in the inset. The rod, with a diameter of 9.2 nm and a length of 48 nm, is capped by two hemispheres in the ends. The dielectric function of Au was taken from JC [38].

2.3.2 Numerical Techniques

The GDM is a semianalytical method because Maxwell's equation is solved analytically by the integral equation Eq. 2.43, while the internal field is solved numerically by replacing the integral equation with discretized algebraic equations. The GDM, however, is restricted to those cases in which the dyadic Green tensor of the background can be constructed. Numerical techniques, such as FDTD and FEM, are usually more convenient among different applications but at the pace of, usually, convergence problems that require particular attention. Detailed descriptions of FDTD and FEM can be found in many excellent textbooks [53, 54] and thus are not presented here.

There are many commercially implemented numerical packages that are capable of handling complex structures. The meshing techniques integrated in these software packages are usually powerful in resolving fine structures, even if the system is large. Figure 2.11 shows an example of the surface charge distribution of six bright modes in a nanocube calculated by a FEM package (COMSOL Multiphysics) [55]. Apparently, the nonuniform tetrahedron-based meshing scheme guarantees the resolution of the highly localized corner modes with good precision. Besides, well-developed software packages are now able to generate structures with significant complexity.

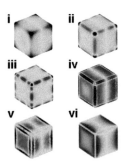

Figure 2.11 FEM-calculated surface charge distribution of six bright modes at (i) 376, (ii) 342, (iii) 330, (iv) 310, (v) 301, and (vi) 292 nm in a 3 nm silver nanocube in vacuum. The incident electric field was polarized along a cube edge. Reprinted with permission from Ref. [56]. Copyright (2011) American Chemical Society.

However, the validity of numerical calculations should be confirmed once a new structure is at hand. For example, enlarging the computation domain and refining the mesh in FDTD and FEM are usually required to ensure the convergence. Unlike Mie theory and the GDM with explicit formulas, pure numerical techniques usually require a postevaluation of the far-field cross sections according to the definitions by Eqs. 2.20 and 2.21. Provided that the model has been correctly set up, numerical simulations can reach an accurate result as those by analytical methods. For comparison, Fig. 2.12 shows the extinction efficiency of a gold sphere immersed in water, calculated by the GDM, FEM, and Mie theory, respectively. In the GDM, the sphere contains $N = 61,565$ subvolumes, corresponding to a mesh of 0.4 nm. In FEM, the computation domain was truncated 230 nm away from the sphere by a 100 nm thick perfectly matched layer and the mesh on the sphere surface is 4 nm, 10 times larger than that in the GDM. Though both GDM and FEM provide a result with a satisfactory overall accuracy compared to Mie theory, FEM has a better agreement with Mie theory with a relative difference smaller than 1.2% over the entire spectrum.

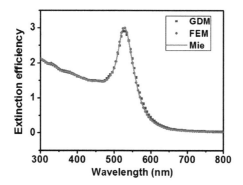

Figure 2.12 Extinction efficiency calculated by the GDM, FEM, and Mie theory for a 40 nm gold sphere embedded in water. The sphere was discretized into N = 61,565 cubic lattices in the GDM. The dielectric function of gold was taken from JC [38].

2.4 Summary

To summarize, we have introduced both analytical and numerical computational techniques for light scattering by small metallic particles. For frequency domain methods, the material response with arbitrary dispersion behaviors can be modeled by a frequency-dependent permittivity or permeability. For a time-domain method like FDTD, a dispersion law should be included. For example, the Drude model works well in describing metals in the near infrared, while certain Lorentzian terms are usually required to include the interband transitions [57]. Although the implements of the computational methods are quite different, they usually end up with the same results. Therefore, the choice of different approaches for a given system is kind of random. For multiple spheres in a homogeneous medium, most of the methods can be used, including generalized Mie theory, the GDM, FDTD, and FEM. For a sphere above an infinitely extended substrate, the GDM and FDTD can be used without additional considerations. However, Mie theory works only for a small particle (compared to the wavelength) when the image particle concept is valid [58]. FEM, however, requires a special treatment on the incident field to model the air–substrate interface [56]. For arbitrarily shaped scatterers in an irregular background, both Mie theory and the GDM fail, so the pure numerical methods should be picked up.

References

1. R. Lord, *Phil. Mag.*, **47**, 375 (1899).
2. G. Mie, *Ann. Phys. (Leipzig)*, **330**, 377 (1908).
3. M. Born and E. Wolf, *Principles of Optics: Electromagnetic Theory of Propagation, Interference and Diffraction of Light*, Seventh (Expanded) edition (Cambridge University Press, 1999).
4. J. A. Stratton, *Electromagnetic Theory* (McGraw-Hill, New York, 1941), p. 349.
5. C. F. Bohren and D. R. Huffman, *Absorption and Scattering of Light by Small Particles* (John Wiley & Sons, New York, 1983).
6. J. J. Sakurai, *Modern Quantum Mechanics* (Addison-Wasley, 1994).
7. M. Inoue and K. Ohtaka, *J. Phys. Soc. Jpn.*, **52**, 3853 (1983).
8. H. X. Xu, *J. Quant. Spectrosc. Radiat. Transfer*, **87**, 53 (2004).
9. S. Stein, *Quart. Appl. Math.*, **19**, 15 (1961).
10. O. R. Cruzan, *Quart. Appl. Math.*, **20**, 33 (1962).
11. C. Liang and Y. T. Lo, *Radio Sci.*, **2**, 1481 (1967).
12. J. H. Bruning and Y. T. Lo, *IEEE Trans. Antennas Propag.*, **AP-19**, 378 (1971).
13. K. A. Fuller, *Appl. Opt.*, **30**, 4716 (1991).
14. G. Gouesbet, *J. Opt. Soc. Am. A*, **16**, 1641 (1999).
15. Z. P. Li, M. Käll, and H. X. Xu, *Phys. Rev. B*, **77**, 085412 (2008).
16. Y. L. Geng, X. B. Wu, L. W. Li, and B. R. Guan, *Phys. Rev. E*, **70**, 056609 (2004).
17. S. Asano and G. Yamamoto, *Appl. Opt.*, **14**, 29 (1975).
18. S. Asano, *Appl. Opt.*, **18**, 712 (1979).
19. A. L. Aden and M. KerKer, *J. Appl. Phys.*, **22**, 1242 (1951).
20. A. Güttler, *Ann. Phys. (Leipzig)*, **11**, 65 (1952).
21. R. Bhandari, *Appl. Opt.*, **24**, 1960 (1985).
22. J. Sinzig and M. Quinten, *Appl. Phys. A*, **58**, 157 (1994).
23. H. X. Xu, PhD thesis, Chalmers University of Technology (2002).
24. H. X. Xu, *Appl. Phys. Lett.*, **85**, 5980 (2004).

25. H. X. Xu, *Phys. Rev. B*, **72**, 073405 (2005).
26. H. X. Xu, *Phys. Lett. A*, **312**, 411 (2003).
27. H. X. Xu, *J. Opt. Soc. Am. A*, **21**, 804 (2004).
28. Z. P. Li and H. X. Xu, *J. Quant. Spectrosc. Radiat. Transfer*, **103**, 394 (2007).
29. S. J. Oldenburg, R. D. Averitt, S. L. Westcott, and N. J. Halas, *Chem. Phys. Lett.*, **288**, 243 (1998).
30. R. Bardhan, S. Lal, A. Joshi, and N. J. Halas, *Acc. Chem. Res.*, **44**, 936 (2011).
31. L. R. Hirsch, R. J. Stafford, J. A. Bankson, S. R. Sershen, B. Rivera, R. E. Price, J. D. Hazle, N. J. Halas, and J. L. West, *Proc. Natl. Acad. Sci. USA*, **100**, 13549 (2003).
32. C. Loo, A. Lowery, N. J. Halas, J. West, and R. Drezek, *Nano Lett.*, **5**, 709 (2005).
33. R. Bardhan, S. Mukherjee, N. A. Mirin, S. D. Levit, P. Nordlander, and N. J. Halas, *J. Phys. Chem. C*, **114**, 7378 (2010).
34. H. X. Xu et al., *Proc. SPIE*, **4258**, 35 (2001).
35. H. X. Xu, E. J. Bjerneld, M. Käll, and L. Borjesson, *Phys. Rev. Lett.*, **83**, 4357 (1999).
36. H. X. Xu, J. Aizpurua, M. Käll, and P. Apell, *Phys. Rev. E*, **62**, 4318 (2000).
37. S. Kim, J. H. Jin, Y. J. Kim, I. Y. Park, Y. Kim, and S. W. Kim, *Nature*, **453**, 757 (2008).
38. P. B. Johnson and R. W. Christy, *Phys. Rev. B*, **6**, 4370 (1972).
39. S. M. Nie and S. R. Emery, *Science*, **275**, 1102 (1997).
40. T. Shegai, Z. P. Li, T. Dadosh, Z. Y. Zhang, H. X. Xu, and G. Haran, *Proc. Natl. Acad. Sci. USA*, **105**, 16448 (2008).
41. K. R. Li, M. I. Stockman, and D. J. Bergman, *Phys. Rev. Lett.*, **91**, 227402 (2003).
42. S. A. Maier, P. G. Kik, H. A. Atwater, S. Meltzer, E. Harel, B. E. Koel, and A. A. G. Requicha, *Nat. Mater.*, **2**, 229 (2003).
43. N. A. Mirin, K. Bao, and P. Nordlander, *J. Phys. Chem. A*, **113**, 4028 (2009).
44. J. B. Lassiter, H. Sobhani, J. A. Fan, J. Kundu, F. Capasso, P. Nordlander, and N. J. Halas, *Nano Lett.*, **10**, 3184 (2010).
45. T. Wriedt, *Part. Part. Syst. Char.*, **15**, 67 (1998).
46. F. M. Kahnert, *J. Quant. Spectrosc. Radiat. Transfer*, **79**, 775 (2003).

47. G. Veronis and S. H. Fan, in *Surface Plasmon Nanophotonics*, edited by M. L. Brongersma, and P. G. Kik (Springer, Berlin, Heidelberg 2007).
48. J. Zhao, A. O. Pinchuk, J. M. Mcmahon, S. Z. Li, L. K. Ausman, A. L. Atkinson, and G. C. Schatz, *Acc. Chem. Res.*, **41**, 1710 (2008).
49. B. Gallinet, J. Butet, and O. J. F. Martin, *Laser Photonics Rev.*, **9**, 577 (2015).
50. P. Gay-Balmaz and O. J. F. Martin, *Appl. Opt.*, **40**, 4562 (2001).
51. M. Paulus, P. Cay-Balmaz, and O. J. F. Martin, *Phys. Rev. E*, **62**, 5797 (2000).
52. B. T. Draine and P. J. Flatau, *J. Opt. Soc. Am. A*, **11**, 1491 (1994).
53. A. Taflove and S. C. Hagness, *Computational Electrodynamics: The Finite-Difference Time-Domain Method,* 3rd ed. (Artech House, Boston, 2005).
54. J. M. Jin, *The Finite Element Method in Electromagnetics,* 2nd ed. (Wiley-IEEE Press, 2002), p. 780.
55. http://www.comsol.com/.
56. S. P. Zhang, K. Bao, N. J. Halas, H. X. Xu, and P. Nordlander, *Nano Lett.*, **11**, 1657 (2011).
57. F. Hao and P. Nordlander, *Chem. Phys. Lett.*, **446**, 115 (2007).
58. H. X. Xu and M. Käll, *Sens. Actuators B*, **87**, 244 (2002).

Chapter 3

Electromagnetic Field Enhancement in Surface-Enhanced Raman Scattering

Ke Zhao,[a] Hong Wei,[a] and Hongxing Xu[b]

[a]*Institute of Physics, Chinese Academy of Sciences, Beijing 100190, China*
[b]*School of Physics and Technology, and Institute for Advanced Studies, Wuhan University, Wuhan 430072, China*
hxxu@whu.edu.cn

3.1 Introduction

Surface-enhanced Raman scattering (SERS) was first observed in 1974 by Fleischmann et al. [1] for pyridine, C_5NH_5, adsorbed on a silver electrode roughened by means of successive oxidation-reduction cycles. This observation of increased Raman intensity was explained by an increase in surface area, that is, an increase in the number of adsorbed molecules contributing to the Raman signal. But, in 1977, Jeanmaire and Van Duyne [2] and, independently, Albrecht and Creighton [3] repeated these experiments and concluded that the enormously strong Raman signal must be caused by a true enhancement of the Raman scattering efficiency itself. Since then

---

Nanophotonics: Manipulating Light with Plasmons
Edited by Hongxing Xu
Copyright © 2018 Pan Stanford Publishing Pte. Ltd.
ISBN 978-981-4774-14-7 (Hardcover), 978-1-315-19661-9 (eBook)
www.panstanford.com

the SERS effect has been observed for many molecules adsorbed on a number of metals [4–6]. On specially prepared rough metal surfaces or colloidal metal nanoparticles, the Raman signal can be a million-fold more intense than that of free molecules. More recently, strikingly large enhancement factors, over 10^{10} in magnitude, have been observed for molecules adsorbed on colloidal silver particles [7–10]. Such a high enhancement factor renders, for the first time, the Raman spectra of a single molecule to be observed.

Combining SERS experiments and theoretically obtained electromagnetic (EM) field distribution around metal surfaces, the EM effect has been recognized as the dominating factor contributing to SERS in most cases, compared to the additional about 10^2 chemical enhancement [4]. When light is incident on the surface of a roughened metal substrate, a local EM field with strength much stronger than the incident light is generated in the vicinity of the substrate. The highly magnified local field significantly strengthens the Raman scattering process of molecules adsorbed on the substrate, which works effectively as an antenna capable of collecting photons. Furthermore, the substrate also enhances the Raman scattering light emitted from the molecules, playing a role of an antenna for broadcasting. Thus the EM effect contributes a total enhancement factor of [9]

$$M_{em} \approx [|\vec{E}_L(\omega_I)|^2/|\vec{E}_I(\omega_I)|^2]\cdot[|\vec{E}_L(\omega_I\pm\omega_v)|^2/|\vec{E}_I(\omega_I\pm\omega_v)|^2], \quad (3.1)$$

where ω_I is the frequency of incident light, ω_v is the vibrational frequency of the molecule undergoing the Raman process, $\vec{E}_I$ is the electric field of the incident light, and $\vec{E}_L$ is the local magnified field feeding the molecule. The first term in the square brackets on the right-hand side of Eq. 3.1 reflects the capability of the substrate to collect incident light, while the second term reflects its capability to enhance the Raman emission from the molecule, where the plus and minus signs in the parentheses correspond to anti-Stokes and Stokes scattering, respectively. Since usually $\omega_v \ll \omega_I$, the EM enhancement factor can be roughly estimated as (abbreviated as E^4)

$$M_{em} \approx |\vec{E}_L(\omega_I)|^4/|\vec{E}_I(\omega_I)|^4. \quad (3.2)$$

The EM effect originates from surface plasmon (SP) excitation, which is the collective electronic excitation mode manifested as coherent electron density oscillation localized at the surface of conductive materials [11, 12]. The surface roughness enables the SP resonance to focus the light at the subwavelength scale, forming hot spots with strong enhancement of the EM field. On the other hand, when the incident light is used to excite the SPs, energies of the photon and plasmon must match, as well as their momenta. These matching conditions may be simultaneously satisfied when the translational symmetry is broken, which renders surface roughness conducive to coupling light to SP resonances.

Extensive studies of EM enhancement have been focused on metal nanoparticles and their aggregates. The localized surface plasmons (LSPs) excited on nanoparticles usually exhibit a large enhancement factor, with the E^4 reaching 10^4 or higher. Using the Drude model [13] and assuming the applicability of quasi-static approximation, we can make the simple rationale: for a nanosphere, its dynamic electric polarizability is $\alpha(\omega) = R^3 \left[1 - \frac{\omega}{\Omega}\left(\frac{\omega}{\Omega} + i\frac{\gamma}{\Omega}\right) \right]^{-1}$, where R is the radius of the sphere, ω and Ω are the frequencies of incident light and LSPs, respectively, and γ measures the effective damping. For incident light with an electric field $\vec{E}_I(\omega)$, the local field around the nanosphere is $\vec{E}_L(\omega) \approx \vec{E}_I(\omega) \cdot \left[1 - \frac{\omega}{\Omega}\left(\frac{\omega}{\Omega} + i\frac{\gamma}{\Omega}\right) \right]^{-1}$.

Since typically the damping is $\gamma \approx 0.1\,\Omega$ or less, E^4 is $\sim 10^4$ or higher at resonance ($\omega = \Omega$).

The most distinct feature of metal nanoparticles is that when they aggregate, extraordinarily high EM enhancement over 10^{10} can be reached in the nanogaps between two particles, rendering single-molecule SERS possible [9, 14, 15]. The extraordinary EM enhancement in nanogaps originates from strong EM coupling between individual nanoparticles, and the coupling strength and enhancement factor rapidly increase when the size of the nanogap decreases. However, when the gap separation goes down to a very small value, smaller than 1 nm, electronic tunneling between the nanoparticles becomes significant, which reduces the interparticle coupling and tends to lower EM enhancement [16].

3.2 Numerical Approaches to EM Enhancement

Because hot spots extend the typical size of only a few nanometers, the EM field within their range cannot be easily probed. Various theoretical and computational approaches capable of evaluating the spatial distribution of EM enhancement around nanostructures with high accuracy have been developed in the last few decades. For nanostructures with all three dimensions larger than a few nanometers, nonlocal effects are negligible and local macroscopic permittivity $\varepsilon(\vec{r},\omega)$ and permeability $\mu(\vec{r},\omega)$ are sufficient for the EM modeling. For these geometries, classical approaches such as finite-difference time-domain (FDTD), generalized Mie theory (GMT), and discrete dipole approximation (DDA) are the most popular ones to simulate the local field distribution [17–19]. If one or more of the dimensions are small, nonlocal response should be considered and in this regard, first-principle linear response theory is an excellent solution [20]. More recently, McMahon et al. incorporated nonlocal response into the fully retarded classical EM simulation by adopting the hydrodynamic model for permittivity, which effectively improved the simulation of local EM enhancement around apexes of nanostructures [21].

The FDTD method discretizes and evolves Maxwell's equations both in space and in time [17]. After propagating the EM wave for a sufficiently long time, Fourier transform is performed to obtain the field distribution in the frequency domain. The discretization and evolution of Maxwell's equations implement the Yee algorithm, for which all the components of the electric field $\vec{E}$ and magnetic field $\vec{H}$ are spatially offset from one another, and time-wise $\vec{E}$ and $\vec{H}$ are mutually offset by a 1/2 time step. Different boundary conditions can be imposed on the simulation region, including the absorbing boundary condition used to model wave-scattering problems and the periodic boundary condition applied to systems containing translational symmetry. FDTD is a universal method applicable to various types of optical response problems, including LSPs of nanoparticles and propagating SPs along nanowires.

GMT is based on Mie theory, which analytically describes the scattering of an EM wave by a single spherical object, and extends to calculate the EM response from aggregates of spheres [18]. Similar to Mie theory for a single sphere, in GMT both incident and scattered fields are expanded in terms of vector spherical harmonics (VSH) $|mnsp\rangle$ serving as the basis. Here m and n are angular momentum quantum numbers; $s = 1$ for the incident wave and 3 for the scattered wave, which satisfies different boundary conditions in faraway space; and $p = 1$ for the transverse electric field $\vec{M}$ and 2 for the transverse magnetic field $\vec{N}$. The great advantage of GMT is that its accuracy does not change with the size of a sphere and it is a highly precise approach to simulate aggregates of spherical nanoparticles. To take account of the interaction between nanospheres, the coordinate transformation for VSH centered at different sites has been developed, which renders a rigorous mathematical connection between incident and scattered waves for different spheres [22].

DDA divides the nanoparticles into small cubes, each of which has its own electric polarizability determined by the local permittivity (usually the material concerned for SERS is nonmagnetic) [19]. The small cubes couple to each other through dipolar interaction, and the electric field radiated from the dipole $\vec{P}_j$ and reaching position $\vec{r}_{ij}$ is

$$\vec{E}_{ij} = k^2 e^{ikr_{ij}} \frac{\vec{r}_{ij} \times (\vec{r}_{ij} \times \vec{P}_j)}{r_{ij}^3} + e^{ikr_{ij}}(1-ikr_{ij})\frac{[r_{ij}^2 \vec{P}_j - 3\vec{r}_{ij}(\vec{r}_{ij} \cdot \vec{P}_j)]}{r_{ij}^5}. \quad (3.3)$$

Eventually one ends up with a set of self-consistent equations for all the dipoles constituting the nanoparticles in the simulated system.

McMahon et al. recently developed a technique to take account of the spatial nonlocal response of materials by incorporating the wave vector $\vec{k}$-dependent permittivity $\varepsilon(\vec{k},\omega)$ into the computational EM methods [21]. This development significantly improves the calculation of EM response from nanostructures consisting of dimensions smaller than 10 nm. In one example, they showed that when nonlocal effects are considered, the field enhancements around the apexes of gold triangular nanowires are significantly lower than the results obtained from local electrodynamics.

A more accurate method for low dimensions, where quantum effects are significant, is to combine linear response theory with electronic structure calculations for the nanostructures [20, 23]. The noninteracting density–density response function is in the following form:

$$\chi(\vec{r},\vec{r}',\omega) = \sum_{i,j}(f_i - f_j)\frac{\psi_i(\vec{r})\psi_j^*(\vec{r})\psi_j(\vec{r}')\psi_i^*(\vec{r}')}{\omega - \omega_{ij} + i\eta}, \quad (3.4)$$

where f_i is the Fermi occupation number for energy ε_i, $\omega_{ij} = (\varepsilon_i - \varepsilon_j)/\hbar$, $\eta \to 0^+$, and Ψ_i is the single-body electronic wave function. The induced charge density that determines the local EM enhancement is calculated as

$$\rho_{\text{ind}}(\vec{r},\omega) = \int \chi(\vec{r},\vec{r}',\omega) \cdot \Phi_{\text{tot}}(\vec{r}',\omega) d^3r', \qquad (3.5)$$

where Φ_{tot} is the distribution of total electric potential. Applying the electrostatic relation $\Phi_{\text{tot}} = \Phi_{\text{ext}} + \Phi_{\text{ind}}$ and $4\pi\rho_{\text{ind}} = -\nabla^2\Phi_{\text{ind}}$, where Φ_{ext} and Φ_{ind} are the electric potentials of the external field and the induced field, respectively, Φ_{ind} and ρ_{ind} can both be self-consistently calculated. The local EM enhancement is calculated as $M = |\nabla\Phi_{\text{tot}}|^4/|\nabla\Phi_{\text{ext}}|^4$.

3.3 The Nanogap Effect

In the year 1997, Nie and Emory [7] and Kneipp et al. [8] reported SERS enhancement as large as 10^{14} in the experiments for single-molecule SERS. However, a clear physical picture for the origin of this large enhancement is absent. Later, Xu et al. concluded that the huge enhancement originates from the nanogaps of aggregates composed of at least two metal nanoparticles, by positioning single nanoparticle structures to correlate the SERS signals and structural morphologies obtained from scanning electron microscopy (SEM) images [9]. Figure 3.1 shows the SEM images of silver (Ag) nanoparticles, which aggregate and form dimers, and the insets of Fig. 3.1c and 3.1d are Raman spectra of single hemoglobin (Hb) molecules. The single-molecule limit is achieved by decreasing and thus limiting the number of molecules adsorbed to each aggregated nanoparticle. Direct evidence for the spectra being at the single-molecule level is their temporal fluctuation, which cannot be the result of the ensemble average over a large number of molecules.

The numerical simulations confirm that the EM enhancement in the nanogaps of nanoparticle dimers can be extraordinarily high and dominate among the factors contributing to single-molecule SERS [9, 14]. Figure 3.2 shows the simulated enhancement spectra for Ag nanosphere dimers and a single nanosphere. As can be seen, the E^4 enhancement factor (M) for the dimer is about 4 orders of magnitude larger than for the single nanosphere when the external electric field is polarized along the dimer axis. The polarization dependence of the enhancement is consistent with the experimental results discussed below. For a nanogap size d = 5.5 nm, approximating the dimension of a Hb molecule, the maximal enhancement (M) is about 10^7 for the wavelengths in experiments. Decreasing the nanogap size from 5.5 to 1 nm results in an increase of M from $\sim 10^7$ to $\sim 10^{10}$. The experimentally estimated enhancement factor of $\sim 10^{10}$ is well within the reach of the EM model without considering the chemical enhancement. It is noted that later studies showing an enhancement factor of 10^7 can be sufficient for single-molecule SERS detection, and the large enhancement factor of 10^{14} claimed in the literature may be an overestimation [24].

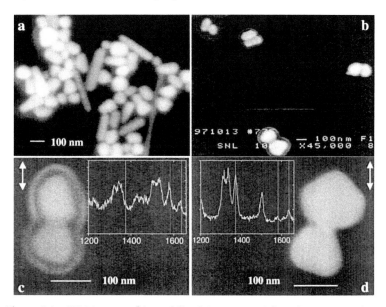

Figure 3.1 SEM images of immobilized Ag nanoparticles. The pictures show (a) an overview of Ag particle shapes and sizes, (b) Ag particle dimers observed after incubation with 1×10^{-11} M Hb molecules for 3 h, and (c, d) hot dimers and corresponding single-Hb-molecule spectra. The double-headed arrows in (c) and (d) indicate the polarization of the incident laser field. Reprinted with permission from Ref. [9]. Copyright (1999) by the American Physical Society.

Figure 3.3 shows the calculated spatial distribution of the E^4 enhancement factor for single Ag nanoparticles (first row) and dimers with different nanogap sizes (second and third rows). The

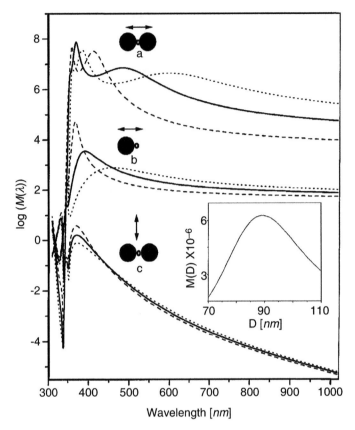

Figure 3.2 Calculated EM enhancement factor for the midpoint between two Ag spheres separated by d = 5.5 nm and for a point $d/2$ outside a single sphere. The solid and open circles indicate the positions of the Ag spheres and the Hb molecule, respectively. The double-headed arrows indicate the incident polarization. The calculations have been performed for spheres of diameters D = 60 (dashed curves), 90 (solid curves), and 120 nm (dotted curves). The inset shows the enhancement versus D for an incident wavelength of 514.5 nm and a Stokes shift of 1500 cm^{-1} for configuration a. Reprinted with permission from Ref. [9]. Copyright (1999) by the American Physical Society.

enhancement in the nanogap of a dimer is orders of magnitude higher than twice of the enhancement near a single nanoparticle because of the strong EM coupling between two nanoparticles. When a pair of closely spaced nanoparticles is excited by incident light, their LSP modes are manifested as electric multipoles and strongly polarize each other, leading to a significantly enhanced induced field compared to that around a single nanoparticle. The mutual polarization is strongest in the nanogap region, which, as a result, contains the highest EM enhancement. The nanoparticles in the right column of Fig. 3.3 are modeled as rotationally symmetric polygons. As can be seen, the introduction of edges to the nanoparticle does not strongly modify either the magnitude of the enhancement or the spatial distribution, which indicates it is reasonable to use nanospheres to model faceted nanoparticles in experiments.

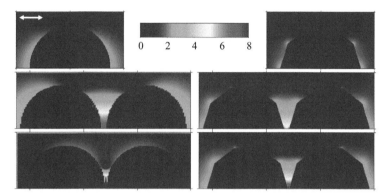

Figure 3.3 EM enhancement factor at a cross section through six different silver particle configurations. The incident field has a wavelength of 514.5 nm with horizontal polarization. The left-hand column shows the EM enhancement for a nanosphere (top) and a dimer of two nanospheres with a separation of 5.5 nm (middle) and 1 nm (bottom). The right-hand column shows the corresponding results for polygons. All particles share a common largest dimension of 90 nm. Note that the color scale from dark blue to dark red is logarithmic, covering the interval $10^0 < M < 10^8$. Regions with enhancement outside this interval are shown in dark blue and dark red. Reprinted with permission from Ref. [14]. Copyright (2000) by the American Physical Society.

The EM enhancement in the nanogap sensitively depends on two factors, the size of the nanogap and the polarization of the external electric field. When the electronic waves of two nanoparticles are well separated, each particle remains neutral during the plasmon excitation. With the decreasing nanogap size, the EM enhancement rapidly increases because of a stronger interparticle coupling (comparing the second and third rows in Fig. 3.3). Figure 3.4 shows the simulated EM enhancement spectra in the nanogaps of model nanosphere dimers and single nanospheres with different geometries. The enhancement in the nanogaps of dimers is several orders of magnitude larger than that of single spheres over the whole visible-to-near-infrared spectral range. For both silver and gold nanospheres, when the nanogap size of the dimers increases from 1 to 5.5 nm, the enhancement dramatically drops by several orders of magnitude over the whole visible-to-near-infrared spectral range.

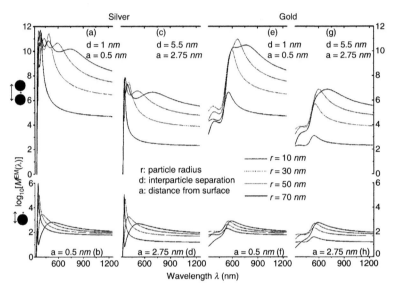

Figure 3.4 The upper row shows the EM enhancement factor M^{EM} versus the wavelength at the midpoint between two equal-radius spherical particles, composed of silver (a, c) and gold (e, g), separated by a gap of d = 1 nm (a, e) and d = 5.5 nm (c, g) for various radii r. When one particle is removed the EM enhancement factor decreases dramatically, as illustrated by the data for single particles shown in the lower row (b, d, f, h). Calculations have been performed for the polarization parallel to the dimer axis and for the surrounding medium of vacuum. The dielectric constants of the particles are based on the experimental data of Johnson and Christy. Reprinted with permission from Ref. [14]. Copyright (2000) by the American Physical Society.

On the other hand, for a specific dimer, the highest EM enhancement is obtained when the external field is polarized along the dimer axis. Therefore, the intensity of the SERS signal is dependent on the laser polarization and the strongest for polarization along the dimer axis [25]. Actually, the enhancement has approximately a $\cos^2\alpha$ dependence on the angle α between the polarization and the dimer axis. This is because the local EM field enhancement in the E^4 law has a $\cos^2\alpha$ polarization dependence, while the Raman emission enhancement is independent of angle α [26]. Figure 3.5 illustrates the experimental results of the polarization dependence of SERS in the dimers, and Fig. 3.6 shows the simulation results of the polarization dependence of EM enhancement [25]. The consistency between Figs. 3.5 and 3.6 again supports that the nanogap effect is the key ingredient of ultrasensitive SERS analysis.

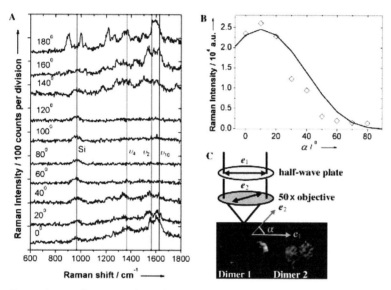

Figure 3.5 Polarization-dependent SERS spectra (A) and SERS intensity (B) from the corresponding dimers shown in the SEM images (C). Dimer 1: 285 × 285 nm² for (A) and Dimer 2: 420 × 420 nm² for (B). The half-wave plate in the inset rotates the polarization vector of the excitation beam (e_1) by an angle α. In (A), the vertical lines mark the heme modes (v_2, v_4, and v_{10}) and the two-phonon band of the Si substrate. In (B), the vertical axis shows the integrated intensity between 700 and 2200 cm^{-1} Stokes shifts; ◊ experimental data, — fit to $\cos^4(\alpha - \alpha_0)$. All measurements were performed using excitation at 514.5 nm and with the same collection time (30 s). The incident irradiance was ~1 µW/µm² in (A) and ~0.1 µW/µm² in (B). The angle α_0 between the dimer axis and the e_1 axis was ~0° in (A) and 10° in (B). Reproduced from Ref. [25], with permission from John Wiley and Sons.

Figures 3.2 and 3.4 also show that the peak positions of EM enhancement in the nanogap of a dimer and around a single nanoparticle are significantly different. The difference originates from EM coupling between nanoparticles, which leads to the hybridization of plasmon modes belonging to individual nanoparticles [27]. As a result of the hybridization, new modes are formed and shifted significantly from the original ones. Figure 3.7 schematically illustrates the plasmon hybridization, where l indexes the angular momentum of the plasmon modes of individual nanoparticles. The plasmon hybridization is in analogy to the combination of atomic orbitals, which forms molecular orbitals with a new set of energy levels.

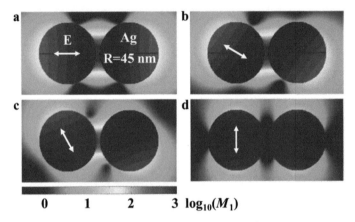

Figure 3.6 Local intensity enhancement $M_1 = (E_{loc}/E_0)^2$ in logarithmic scale in a plane through the centers of the Ag nanospheres and perpendicular to the incident wave vector for different incident polarizations indicated by the arrows: (a) 0°, (b) 30°, (c) 60°, and (d) 90°. Reproduced from Ref. [25], with permission from John Wiley and Sons.

The hybridization of plasmon modes will lead to the change of plasmon resonance frequencies that can be observed from extinction spectra. Figure 3.8 shows the calculated EM enhancement spectra and extinction/scattering/absorption spectra for Ag nanoparticle dimers and pentamers [18]. For the dimer, the long-wavelength peak from the coupling of the dipoles of the two particles is prominent in both the EM enhancement spectra of the nanogap site and the extinction spectra for small polarization angles (0° and 30°). The short-wavelength peaks in the extinction spectra for 0° and 30° are mainly from quadrupolar resonances. When the polarization is perpendicular to the dimer axis, the coupled dipolar mode shifts to short wavelength with low strength, and the EM enhancement in the nanogap more or less vanishes. The accordance between the near-field EM enhancement and far-field extinction can be seen from Fig. 3.8a. For the pentamer, however, the results are quite different, as shown in Fig. 3.8b. The extinction spectra are dominated by a broad long-wavelength peak of dipolar origin. In contrast, the near-field intensities in the nanogaps exhibit resonances much sharper than for the dimer and are extremely polarization dependent. The narrower resonances are expected to originate from the coherent superposition of the EM field from different particles. The results for the pentamer indicate that the properties of near field and far field for the same aggregate can be distinctly different. The numerous nanogaps in the multiple-particle aggregate may become "hot" for different wavelengths and different polarizations. It is noted that the resonance peaks in the spectra show asymmetric profiles, which originate from the interference of dipolar and high-order modes, that is, the so-called Fano resonances.

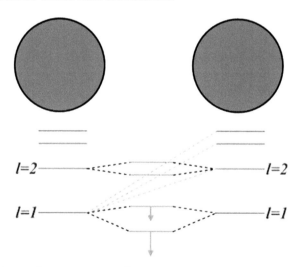

Figure 3.7 Schematic picture illustrating the plasmon hybridization in a nanoparticle dimer. The individual nanosphere plasmons on the two particles interact and form bonding and antibonding dimer plasmons. In the dimer geometry, nanosphere plasmons with a given angular momentum l interact also with plasmons of a different angular momentum on the other particle. This interaction induces extra shifts of the dimer plasmons at small separations. The figure illustrates this shift for the $l = 1$ derived dimer plasmons. Reprinted with permission from Ref. [27]. Copyright (2004) American Chemical Society.

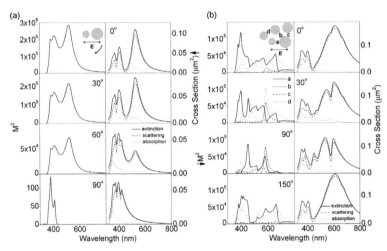

Figure 3.8 Near-field intensity enhancement $M^2(\lambda)$ at hot gap sites and far-field cross sections (extinction, scattering, absorption) for a dimer (diameter $D = 60$ nm and 100 nm) and a pentamer ($D = 60$ nm, 70 nm, 80 nm, 90 nm, and 100 nm) in vacuum with gap dimensions $d = 1$ nm throughout. The spheres are situated on a planar virtual surface, and the incident wave vector is normal to this surface, with polarization as indicated in the insets. The calculations included multipoles up to order $N = 30$ and scattering processes up to infinite order through the matrix inversion technique. Reproduced from Ref. [18]. With permission of Springer.

3.4 Various Types of Nanogaps

In addition to aggregates of nanoparticles, nanogaps favored by Raman scattering are also formed in several other types of systems. One example is tip-enhanced Raman scattering (TERS), with the molecules trapped in the nanogap between a metal tip and a metal substrate [28–34]. The configuration is schematically illustrated in Fig. 3.9 (molecule not shown) [35]. According to the numerical simulation carried out by Yang et al., when the distance between the gold tip and the gold substrate is $d = 2$ nm, their strong near-field EM coupling results in 1.8×10^9 EM enhancement for TERS in the nanogap. The coupling strength in TERS is also strongly dependent on the tip–substrate separation since the EM near field decays in an exponential manner far from the surface. On the other hand, a sharp tip with a small radius produces a relatively large enhancement and high spatial resolution of TERS, which is due to the lightning rod effect resulting from the increasing confinement of the surface charges at the sharp tip apex. Figure 3.10 shows the simulated maximum EM enhancement in the nanogap for different tip radii and the spatial distribution of enhancement for a gold tip with a radius of 25 nm. When the radius decreases from 10 to 5 nm, the enhancement in the nanogap is rapidly increased. The spatial resolution quantified as the full-width at half-maximum (FWHM) of the Raman enhancement is less than 10 nm for the 25 nm radius gold tip.

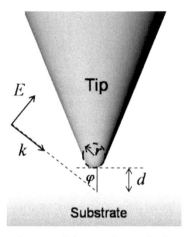

Figure 3.9 Schematic diagram of the metal tip and substrate geometry: the tip is modeled as a conical taper terminated by a hemisphere of radius r, held at a distance d from the substrate surface. An EM plane wave is incident at an angle φ, with the polarization in the plane of incidence. Reproduced from Ref. [35], with permission from John Wiley and Sons.

Another type of nanogap is formed between a metal nanoparticle and a nanowire [26, 36], and the particle–wire coupling results in a strong EM enhancement, showing a SERS signal comparable to those in the nanogaps of nanoparticle dimers. Figure 3.11a shows the experimental SERS spectra of malachite green isothiocyanate (MGITC) molecules at different spots of the coupled system [26]. Clearly the highest enhancement is in the junction between the nanowire and the nanoparticle. The EM simulation shows that for the 5 nm gap separation between the particle and wire, the average SERS enhancement factor is of the order of 10^6 for the different shapes of particles and the maximum local SERS enhancement for perpendicular polarization is of the order of 10^{10}, which is similar to the maximum enhancement in the nanogap of nanoparticle dimers. The polarization dependence of the EM enhancement in the nanogap can be clearly seen from Fig. 3.11b.

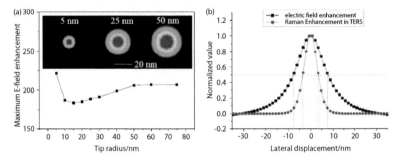

Figure 3.10 Tip size dependence of the maximum field enhancement and spatial resolution for a gold tip held at 2 nm above a gold substrate surface. The 632.8 nm plane wave is incident from the side at an angle of 60° with respect to the substrate normal, with the polarization in the plane of incidence. (a) Maximum electric field enhancement in the nanogap as a function of the tip radius. The insets show the spatial field distribution in a plane parallel to and 1 nm above the substrate for tip radius r = 5, 25, and 50 nm, from left to right. (b) Normalized electric field enhancement M and Raman enhancement M^4 along a horizontal line 1 nm below the apex of a gold tip with a 25 nm radius above a gold substrate. Reproduced from Ref. [35], with permission from John Wiley and Sons.

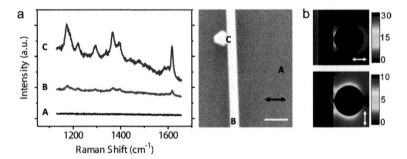

Figure 3.11 (a) Raman spectra of MGITC from different positions of the sample. The arrow in the SEM image shows the incident polarization. The scale bar is 400 nm. (b) Calculated electric field for a gold sphere of radius 50 nm at 5 nm from a wire of radius 25 nm for polarization perpendicular (top) and parallel (bottom) to the wire. Reprinted with permission from Ref. [26]. Copyright (2008) American Chemical Society.

For an infinitely long metal nanowire, which is the ideal model for a long enough wire in reality, the plasmons are continuous modes characterized by wave vectors along the axis and their azimuthal symmetry. When a finite metal nanoparticle is placed in the proximity of the nanowire, the discrete plasmon modes of the particle hybridize with the wire plasmon continuum through EM interaction, which is schematically illustrated in Fig. 3.12 [26]. The interaction results in a set of bonding plasmonic virtual states (VS) and antibonding localized states. The VS are primarily composed of long-wavelength wire plasmons and are optically active because of the finite admixture of dipolar nanoparticle plasmons. The nanoparticle serves as a nanoantenna that when polarized by the incident light couples efficiently to wire plasmons of half wavelengths larger than the diameter of the nanoparticle [37]. The wave vector distribution of wire plasmons that make up the VS does not depend on the detailed shape of the nanoparticle but only on its lateral dimension along the wire. The energies of the wire plasmons that participate in the VS are determined by the wire plasmon dispersion relations, which are dependent on the radius of the wire. The composition and spectral properties of the plasmonic VS are thus determined by the nanoparticle and wire diameters rather than by the detailed geometric structure on the nanoparticle. As a result, the EM enhancement is also remarkably insensitive to the detailed geometry. Moreover, the resonant EM enhancement in the nanogap is highest when the external electric field is polarized perpendicular to the nanowire and across the nanogap, because of the strongest EM coupling between the two entities.

The particle–wire geometry also enables remote excitation SERS [36, 38]. The illumination of one end of the nanowire with a laser can launch the SPs in the wire, which propagate to the site of the nanoparticle and excite its plasmons, leading to a strong coupling between the particle and the wire. The high EM enhancement in the nanogap between the nanowire and the nanoparticle renders single-molecule SERS to be achieved remotely from the laser illumination point. Moreover, when several nanoparticles are attached to one nanowire, multisite remote excitation SERS can be realized in multiple particle–wire nanogaps.

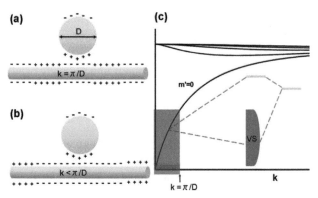

Figure 3.12 Schematic illustration of the coupling of a nanoparticle to the symmetric ($m' = 0$) wire plasmon continuum. (a, b) The nanoparticle can couple efficiently to symmetric wire plasmons of half wavelengths larger than the diameter D. (c) Plasmon hybridization in the metal nanoparticle–nanowire system. The red area illustrates the wave vectors of the wire plasmons that couple efficiently to the nanoparticle, and the green area shows their energies. The VS continuum can couple efficiently to light because of the admixture of nanoparticle plasmons with dipolar moments. Reprinted with permission from Ref. [26]. Copyright (2008) American Chemical Society.

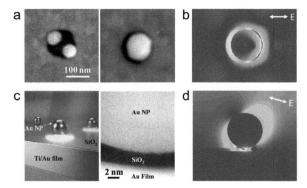

Figure 3.13 (a) SEM images of Au nanoparticle–nanohole structures. (b) Calculated spatial distribution of the electric field for a coupled nanoparticle–nanohole with an excitation wavelength of 633 nm. The diameters of the particle and hole are 100 nm and 120 nm, respectively. The smallest separation between them is 3 nm. (c, left) Schematic diagram of a nanoparticle–film structure; (c, right) high-resolution TEM image of an individual Au nanoparticle on a 2 nm SiO_2 spacer, showing its clear separation from the underlying Au film. (d) Calculated spatial distribution of the electric field with an excitation wavelength of 633 nm. The diameter of the nanoparticle is 35 nm, and the thickness of the SiO_2 layer is 3 nm. (a) Reproduced from Ref. [39] with permission from The Royal Society of Chemistry. (b) Adapted from Ref. [40], with permission from John Wiley and Sons. (c, d) Reprinted with permission from Ref. [41]. Copyright (2012) American Chemical Society.

By putting a metal nanoparticle into a nanohole in metal film, a nanogap can be formed between them [40]. The magnitude of the EM enhancement in the particle–hole nanogap is similar to a nanoparticle dimer. But the volume with large enhancement accessible to molecules is much larger for the particle–hole pair than for the particle dimer. A nanogap can also be created by depositing a nanoparticle onto a metal film with a very thin spacer between them [41]. The large area of the metal film makes it easy to prepare this kind of nanogap. Figure 3.13 shows the SEM/transmission electron microscopy (TEM) images of nanoparticle–nanohole and nanoparticle–film structures and the calculated electric field distributions. It can be clearly seen that the hot spots with large electric field enhancement are located within the nanogaps.

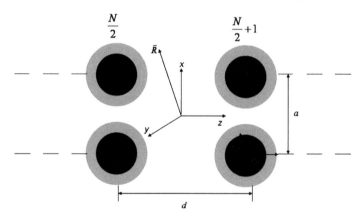

Figure 3.14 Geometrical illustration of an ordered 1D array of metal nanoshell dimers. The metal shell is silver, and the dielectric core is silica. Reprinted from Ref. [43], with the permission of AIP Publishing.

Nanoparticles can be aligned to form ordered arrays. In these arrays, collective photonic effects arising from the long-range order are superposed onto the plasmon excitation of individual array units, such as nanoparticles or dimers. Theoretical modeling and analysis have shown that photonic effects may further increase the EM enhancement of individual units, when their separations are comparable to the plasmon resonance wavelength [42, 43]. Figure 3.14 shows the structure of modeled 1D array of dimers composed of nanoshells, which are metal shells supported by dielectric cores. When a whole dimer array is excited by the plane wave, plasmon modes of dimers are excited and consequently long-range dipolar fields are radiated. The dipolar interaction between dimers enhances the effective field experienced by individual dimers compared to the external field, rendering an additional enhancement in the nanogaps of the dimers. Thanks to the E^4 law for SERS, the additional enhancement can reach up to 1 order of magnitude, as shown by Fig. 3.15.

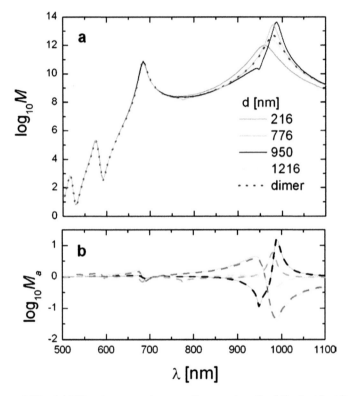

Figure 3.15 (a) EM enhancement versus the wavelength of the incident light at the location halfway between the two centers of a metal nanoshell dimer in dimer arrays of different interdimer distances d. The inner and outer radii of the nanoshell are 35 nm and 38 nm, respectively. The separation distance between the two nanoshells in a dimer is 1 nm. The dotted line is for an isolated dimer of identical geometry. All the curves are calculated by GMT. (b) The solid lines are the additional enhancements due to long-range photonic effects, given by the subtractions of the dotted line in (a) from the solid lines in (a). The dashed lines are the results from the dipole approximation, which takes account of the interdimer interaction only by dipolar interaction. Reprinted from Ref. [43], with the permission of AIP Publishing.

Several different types of ordered nanoparticle arrays have been fabricated and used for SERS [44–47]. In particular, Gunnarsson et al. fabricated highly uniform arrays of nanogaps using electron beam lithography (EBL) to study the interparticle coupling effects for SERS, with the SEM images of the arrays shown in Fig. 3.16 [44]. EBL has the advantage of being able to delicately control the dimensions of the systems and the coupling between nanoparticles. They found that the dependence of SERS intensity on interparticle separation d can be quite well fitted by $I^{SERS} \propto (d+D)^{-2}[A(D/d+1)^4 +B]$, where D is the particle length scale and A and B are fitting parameters. Sawai et al. prepared silver dimer arrays using the nanosphere lithography technique and detected SERS spectra from a small number of molecules in the nanogaps of the dimers [45]. The highly uniform and controllable nanogap arrays may be useful for large-scale SERS substrates or other optical applications.

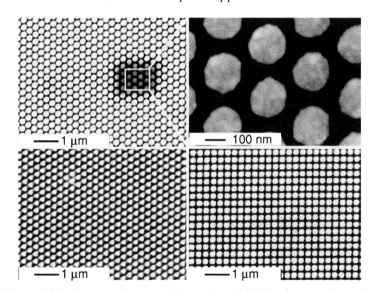

Figure 3.16 Scanning electron micrographs of SERS substrates. Examples consist of circular (top left and right), triangular (bottom left), and square (bottom right) 30 nm thin Ag particles on Si. The predefined particle length scale, defined as the diameter in the case of circular particles and the edge length in the case of triangles or squares, was D = 200 nm, and the predefined interparticle separation distance, defined as the minimum edge-to-edge distance, was d = 100 nm. Reprinted from Ref. [44], with the permission of AIP Publishing.

3.5 Multiple-Particle Nanoantennas for Controlling Polarization of SERS Emission

The nanogap in aggregated nanoparticles also provides another dimension to tune the properties of SERS by manipulating the polarization of light emitted from molecules trapped in the gaps [48, 49]. To significantly rotate the polarization, one or more nanoparticles must be added to the dimer in order to break the dipolar symmetry of nanoparticle aggregates. The simplest aggregate capable of modulating the polarization is a trimer composed of three nanoparticles, whose asymmetry results in a wavelength-dependent rotation of polarization of the Raman scattering light. Figure 3.17 shows the rotation of polarization calculated by GMT for the trimer, with the third particle placed at different positions relative to the dimer [48]. To quantify the rotation, the depolarization ratio is used:

$$\rho(\theta) = \frac{I_x(\theta) - I_y(\theta)}{I_x(\theta) + I_y(\theta)}, \quad (3.6)$$

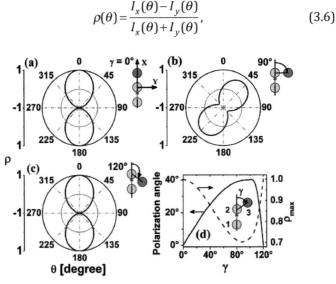

Figure 3.17 Depolarization ratio of the dipole emission from a Ag trimer antenna with (a) linear, (b) right-angle, and (c) equilateral triangle configurations. (d) Polarization angle (left axis) and the maximum depolarization ratio ρ_{max} (right axis) of the antenna emission as a function of angle γ. The radii of all three Ag nanoparticles are identical (R = 40 nm). The separations between the first and second nanoparticles and the second and third nanoparticles are kept at 1 nm. The wavelength of the dipole emission is 555 nm. The refractive index of the surrounding medium is n_s = 1. Reprinted with permission from Ref. [48]. Copyright (2009) American Chemical Society.

where I_x and I_y are the components of the emission projected on the x-y coordinate system. Figure 3.17b shows that the breaking of symmetry may not only rotate the polarization of scattered light but also make it elliptical as characterized by a maximum depolarization ratio ρ_{max} smaller than 1. The ellipticity of the emission is defined as

$$T = (1-\rho_{max})/(1+\rho_{max}). \tag{3.7}$$

The ellipticity in Fig. 3.17b is T = 0.18. The experimentally measured SERS emission from the single molecule in the nanogap of a Ag nanoparticle trimer shows wavelength-dependent polarization rotation and ellipticity, which agrees well with the simulation results. Further simulations show that the polarization rotation is also dependent on the distance between the third particle and the dimer and the size of the third particle [49].

3.6 Electronic Coupling in Nanogaps

When nanoparticles are so close that they nearly touch each other, their electron orbitals may overlap and hybridize, which strongly modifies the electronic structure of the aggregates and thus influences their optical properties [16, 23, 50]. One consequence of the nearly touching limit is a dramatic blue shift of plasmon energy of the aggregates when the interparticle separation decreases to the conductively coupled regime [51–53]. This effect has been demonstrated by the measured optical transmission spectra of an array of gold nanodisk dimers, as shown in Fig. 3.18 [51]. The valleys of spectra correspond to the LSP energy of dimers (long-range interdimer interaction has little effect on the LSP energy) because incident light with those frequencies is absorbed and converted into plasmon excitation. When the electron orbitals of individual nanodisks are distinctly separated from each other and the electronic coupling between disks is weak, the LSP energy red-shifts with the decreasing gap size as a result of plasmon hybridization. But at the nearly touching regime, the LSP energy suddenly displays a significant blue shift due to strong electronic coupling between nanodisks and a substantially modified electronic structure of the whole dimer.

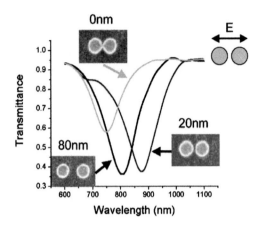

Figure 3.18 Polarized transmission spectra in periodic arrays of pairwise interacting gold nanoparticles. The lattice constant is 800 nm in the parallel direction to the pair axis and 400 nm in the perpendicular direction, and the dot height is 30 nm. The separation distance between the two particles of a pair is labeled above the SEM images. Reprinted with permission from Ref. [51]. Copyright (2004) American Chemical Society.

Another consequence of the strong electronic coupling is that electrons may tunnel across the dimer junction, significantly modifying the optical response and reducing the EM enhancement relative to the classical predictions. Quantum mechanical calculations indicate that this tunneling effect is important for a gap size smaller than ~1 nm [16, 23], which may still hold small molecules. For larger gap sizes, the results of quantum mechanical and classical calculations gradually merge, further confirming the validity of applying classical EM theory to aggregates with wide enough gaps. Using time-dependent local density approximation (TDLDA), Zuloaga et al. studied EM enhancement in the nanogap between two electronically coupled jellium nanospheres and explicitly demonstrated their difference from results obtained by classical theory [23]. As shown in Fig. 3.19, three distinct regimes of interaction between nanoparticles are identified: the classical regime (Figs. 3.19a-i and 3.19b-A), where the gap separation d is larger than ~1 nm and interparticle interaction can be well characterized by plasmon hybridization; the crossover regime (Figs. 3.19a-ii, 3.19b-B, and 3.19b-C), where electrons start to tunnel through a narrow barrier, reducing the EM coupling between the two particles and consequently EM enhancement in the nanogap; and the conductive regime (Figs. 3.19a-iii and 3.19b-D), where the Fermi level of the system lies above the electron potential barrier separating the two particles and the conductance of the gap is large. The enhancement in the crossover and conductive (touching) regimes clearly differentiates quantum mechanical and classical predictions (Fig. 3.19b).

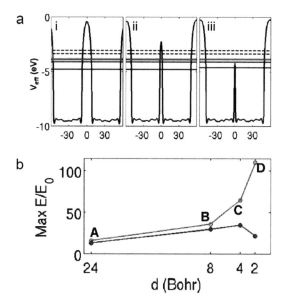

Figure 3.19 (a) Self-consistent effective potential along the interparticle axis of the dimer for a nanosphere of radius $R = 24b$ and separation $d = 16b$ (i), $5b$ (ii), and $2b$ (iii). Here b is Bohr's radius, which is 0.0529 nm. The top horizontal, solid lines mark the Fermi level. (b) Comparison of the maximum electric field enhancements calculated using plasmon hybridization (upper) and TDLDA (lower) for $R = 24b$ dimers of different separation d. Reprinted with permission from Ref. [23]. Copyright (2009) American Chemical Society.

3.7 Probing EM Enhancement via SERS

Because of the small scale of hot spots, using normal sensors to measure EM field distribution is generally not feasible. Yet the EM effect should enable molecules to effectively probe the field distribution by measuring their SERS signal, as long as the chemical effect can be suppressed. Wang et al. combined simulations and experiments to determine the field distribution by studying the SERS of rhodamine 6G (R6G) molecules as they diffuse into Ag@SiO$_2$ core-shell nanoparticles [54]. The Ag@SiO$_2$ nanoparticle is composed of a SiO$_2$ shell supported by a Ag core. When R6G diffuses in the confined SiO$_2$ nanoporous structure and does not chemically bind to Ag, SERS should purely originate from EM enhancement, thanks to LSPs of the Ag core. Mie theory is used to simulate the field distribution in the SiO$_2$ shell, and diffusion theory is applied to calculate the evolution of the R6G concentration in the shell. The concentration of R6G within the SiO$_2$ shell $C(r, t)$ is obtained by solving the equation

$$\frac{\partial C(r,t)}{\partial t} = D \frac{1}{r^2} \frac{\partial}{\partial r}\left[r^2 \frac{\partial C(r,t)}{\partial r}\right], \tag{3.8}$$

where D is the effective diffusion coefficient, t is the time, and r is the radial coordinate. D is the only adjustable parameter used in calculations. The average SERS enhancement factor G_{ave} is obtained by integrating all Raman scattering contributions from the volume of the spherical shell (v):

$$G_{\text{ave}}(t) = \int_v G_{\text{loc}}(r)C(r,t)dV, \tag{3.9}$$

where G_{loc} is the local enhancement at spot r in the shell. Theoretical values of average enhancement $G_{\text{ave}}(t)$ in the SiO$_2$ shell are thus obtained and compared to experimental measurement. The good agreement between theoretical and experimental results in Fig. 3.20a is a solid proof that the calculated field distribution is an excellent simulation of the true distribution. Figure 3.20b shows the EM field distribution quantified by local enhancement.

By measuring near-field Raman scattering and near-field two-photon-induced photoluminescence using an apertured scanning near-field optical microscope, Imura et al. demonstrated that the large EM enhancement in a dimer of gold nanoparticles is localized in the nanogap between the two particles when the laser light is polarized parallel to the interparticle axis [55]. The hot spots in the nanogaps can be experimentally mapped by using the super-resolution imaging technique. Willets et al. mapped the single-molecule SERS hot spot by point spread function fitting, and a resolution of the order of ~5 nm was achieved [56]. The super-resolution optical images reveal the size and field distribution of the hot spots in nanogaps of aggregated nanoparticles. These results provide direct experimental evidence for the large EM enhancement in the nanogaps and its dominant contribution to single-molecule SERS.

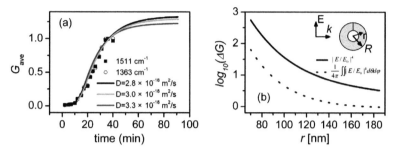

Figure 3.20 (a) Comparison between experimentally observed (solid squares and open circles) and theoretically predicted (lines) SERS enhancement factor, G_{ave}, using spectral data at 1511 or 1363 cm^{-1}. All data were normalized to the saturation value collected at 40 min. The estimated effective diffusion coefficient, D, is 2.8×10^{-18} to 3.3×10^{-18} m^2/s. (b) Local EM intensity enhancement (ΔG) along the transect of the silica shell shown in the inset (solid line) and the average ΔG over the surface at radius r (dashed line). Reprinted with permission from Ref. [54]. Copyright (2009) American Chemical Society.

3.8 Summary

Combining EM theory and SERS experiments, the EM effect has been determined to be the dominating factor for most SERS systems. The EM effect originates from SP excitation and obeys the E^4 law. EM effect is extraordinarily high in the nanogaps of nanoparticle aggregates, making single-molecule SERS possible. Various types of nanogaps have been realized in experiments, including those between nanoparticles, between a tip and a substrate (TERS), between a nanoparticle and a nanowire, etc. In particular, the gap in the particle–wire geometry enables the multisite remote excitation for SERS. When the nanoparticles form ordered arrays, the long-range photonic effect is coupled to LSPs of individual array units, which may further improve the SERS sensitivity. When more than two nanoparticles are aggregated, the third particle may break the dipolar symmetry and rotate the polarization of Raman emission.

Electronic coupling between nanoparticles becomes significant when the particles nearly touch each other. In the conductive regime, the electron orbitals of the particles overlap and hybridize, rendering electrons to tunnel between the particles and lowering the EM enhancement in the nanogap. Finally, probing EM field distributions around nanoparticles and nanogaps can be achieved by delicate designs of the SERS experiments.

References

1. M. Fleischmann, P. J. Hendra, and A. J. McQuillan, *Chem. Phys. Lett.*, **26**, 163–166 (1974).
2. D. L. Jeanmaire and R. P. Van Duyne, *J. Electroanal. Chem.*, **84**, 1–20 (1977).
3. M. G. Albrecht and J. A. Creighton, *J. Am. Chem. Soc.*, **99**, 5215–5217 (1977).
4. M. Moskovits, *Rev. Mod. Phys.*, **57**, 783–826 (1985).
5. A. Otto, Surface-enhanced Raman-scattering: "Classical" and "Chemical" origins. M. Cardona and G. Guntherodt, eds., *Light Scattering in Solids IV* (Springer-Verlag, 1984), pp. 289–418.
6. Z. Q. Tian, B. Ren, and D. Y. Wu, *J. Phys. Chem. B*, **106**, 9463–9483 (2002).
7. S. M. Nie and S. R. Emory, *Science*, **275**, 1102–1106 (1997).
8. K. Kneipp, Y. Wang, H. Kneipp, L. T. Perelman, I. Itzkan, R. Dasari, and M.S. Feld, *Phys. Rev. Lett.*, **78**, 1667–1670 (1997).
9. H. X. Xu, E. J. Bjerneld, M. Käll, and L. Borjesson, *Phys. Rev. Lett.*, **83**, 4357–4360 (1999).
10. A. M. Michaels, J. Jiang, and L. Brus, *J. Phys. Chem. B*, **104**, 11965–11971 (2000).
11. J. M. Pitarke, V. M. Silkin, E. V. Chulkov, and P. M. Echenique, *Rep. Prog. Phys.*, **70**, 1–87 (2007).
12. W. L. Barnes, A. Dereux, and T. W. Ebbesen, *Nature*, **424**, 824–830 (2003).
13. M. P. Marder, *Condensed Matter Physics*, 1st ed. (Wiley, New York, 2000).
14. H. X. Xu, J. Aizpurua, M. Käll, and P. Apell, *Phys. Rev. E*, **62**, 4318–4324 (2000).
15. H. X. Xu, *Appl. Phys. Lett.*, **85**, 5980–5982 (2004).

16. L. Mao, Z. P. Li, B. Wu, and H. X. Xu, *Appl. Phys. Lett.*, **94**, 243102 (2009).
17. A. Taove, and S. C. Hagness, *Computational Electrodynamics: The Finite-Difference Time-Domain Method,* 3rd ed. (Artech House, 2005).
18. H. X. Xu, and M. Käll, Estimating SERS properties of silver-particle aggregates through generalized Mie theory. K. Kneipp, M. Moskovits and H. Kneipp, eds., *Surface-Enhanced Raman Scattering: Physics and Applications* (Springer, 2006), pp. 87–104.
19. W. H. Yang, G. C. Schatz, and R. P. Van Duyne, *J. Chem. Phys.*, **103**, 869–875 (1995).
20. I. Grigorenko, S. Haas, and A. F. J. Levi, *Phys. Rev. Lett.*, **97**, 036806 (2006).
21. J. M. McMahon, S. K. Gray, and G. C. Schatz, *Phys. Rev. Lett.*, **103**, 097403 (2009).
22. O. R. Cruzan, *Quart. Appl. Math.*, **20**, 33 (1962).
23. J. Zuloaga, E. Prodan, and P. Nordlander, *Nano Lett.*, **9**, 887–891 (2009).
24. P. G. Etchegoin, E. C. Le Ru, and M. Meyer, *Phys. Today*, **61**, 13–14 (2008).
25. H. X. Xu and M. Käll, *ChemPhysChem*, **4**, 1001–1005 (2003).
26. H. Wei, F. Hao, Y. Z. Huang, W. Z. Wang, P. Nordlander, and H. X. Xu, *Nano Lett.*, **8**, 2497–2502 (2008).
27. P. Nordlander, C. Oubre, E. Prodan, K. Li, and M. I. Stockman, *Nano Lett.*, **4**, 899–903 (2004).
28. N. Hayazawa, Y. Inouye, Z. Sekkat, and S. Kawata, *Opt. Commun.*, **183**, 333–336 (2000).
29. J. Steidtner and B. Pettinger, *Phys. Rev. Lett.*, **100**, 236101 (2008).
30. J. N. Chen, W. S. Yang, K. Dick, K. Deppert, H. Q. Xu, L. Samuelson, and H. X. Xu, *Appl. Phys. Lett.*, **92**, 093110 (2008).
31. Z. Liu, S. Y. Ding, Z. B. Chen, X. Wang, J. H. Tian, J. R. Anema, X. S. Zhou, D. Y. Wu, B. W. Mao, X. Xu, B. Ren, and Z. Q. Tian, *Nat. Commun.*, **2**, 305 (2011).
32. M. T. Sun, Z. L. Zhang, H. R. Zheng, and H. X. Xu, *Sci. Rep.*, **2**, 647 (2012).
33. E. M. van Schrojenstein Lantman, T. Deckert-Gaudig, A. J. G. Mank, V. Deckert, and B. M. Weckhuysen, *Nat. Nanotechnol.*, **7**, 583–586 (2012).
34. R. Zhang, Y. Zhang, Z. C. Dong, S. Jiang, C. Zhang, L. G. Chen, L. Zhang, Y. Liao, J. Aizpurua, Y. Luo, J. L. Yang, and J. G. Hou, *Nature,* **498**, 82–86 (2013).
35. Z. L. Yang, J. Aizpurua, and H. X. Xu, *J. Raman Spectrosc.*, **40**, 1343–1348 (2009).

36. Y. R. Fang, H. Wei, F. Hao, P. Nordlander, and H. X. Xu, *Nano Lett.*, **9**, 2049–2053 (2009).
37. F. Hao and P. Nordlander, *Appl. Phys. Lett.*, **89**, 103101 (2006).
38. J. A. Hutchison, S. P. Centeno, H. Odaka, H. Fukumura, J. Hofkens, and H. Uji-i, *Nano Lett.*, **9**, 995–1001 (2009).
39. H. Wei and H. X. Xu, *Nanoscale*, **5**, 10794–10805 (2013).
40. H. Wei, U. Hakanson, Z. L. Yang, F. Hook, and H. X. Xu, *Small*, **4**, 1296–1300 (2008).
41. S. Mubeen, S. P. Zhang, N. Kim, S. Lee, S. Kramer, H. X. Xu, and M. Moskovits, *Nano Lett.*, **12**, 2088–2094 (2012).
42. S. L. Zou and G. C. Schatz, *Chem. Phys. Lett.*, **403**, 62–67 (2005).
43. K. Zhao, H. X. Xu, B. H. Gu, and Z. Y. Zhang, *J. Chem. Phys.*, **125**, 081102 (2006).
44. L. Gunnarsson, E. J. Bjerneld, H. X. Xu, S. Petronis, B. Kasemo, and M. Käll, *Appl. Phys. Lett.*, **78**, 802–804 (2001).
45. Y. Sawai, B. Takimoto, H. Nabika, K. Ajito, and K. Murakoshi, *J. Am. Chem. Soc.*, **129**, 1658–1662 (2007).
46. C. M. Ruan, G. Eres, W. Wang, Z. Y. Zhang, and B. H. Gu, *Langmuir*, **23**, 5757–5760 (2007).
47. L. D. Qin, S. L. Zou, C. Xue, A. Atkinson, G. C. Schatz, and C. A. Mirkin, *Proc. Natl. Acad. Sci. USA*, **103**, 13300–13303 (2006).
48. Z. P. Li, T. Shegai, G. Haran, and H. X. Xu, *ACS Nano*, **3**, 637–642 (2009).
49. T. Shegai, Z. P. Li, T. Dadosh, Z. Y. Zhang, H. X. Xu, and G. Haran, *Proc. Natl. Acad. Sci. USA*, **105**, 16448–16453 (2008).
50. M. C. Troparevsky, K. Zhao, D. Xiao, A. G. Eguiluz, and Z. Y. Zhang, *Phys. Rev. B*, **82**, 045413 (2010).
51. T. Atay, J. H. Song, and A. V. Nurmikko, *Nano Lett.*, **4**, 1627–1631 (2004).
52. R. Esteban, A. G. Borisov, P. Nordlander, and J. Aizpurua, *Nat. Commun.*, **3**, 825 (2012).
53. K. J. Savage, M. M. Hawkeye, R. Esteban, A. G. Borisov, J. Aizpurua, and J. J. Baumberg, *Nature*, **491**, 574–577 (2012).
54. W. Wang, Z. P. Li, B. H. Gu, Z. Y. Zhang, and H. X. Xu, *ACS Nano*, **3**, 3493–3496 (2009).
55. K. Imura, H. Okamoto, M. K. Hossain, and M. Kitajima, *Nano Lett.*, **6**, 2173–2176 (2006).
56. M. L. Weber and K. A. Willets, *J. Phys. Chem. Lett.*, **2**, 1766–1770 (2011).

Chapter 4

Plasmonic Antennas

Zhipeng Li,[a] Longkun Yang,[a] Hancong Wang,[a] Pan Li,[a] and Hongxing Xu[b]

[a]*The Beijing Key Laboratory for Nano-Photonics and Nano-Structure (NPNS), Department of Physics, Capital Normal University, Beijing 100048, PR China*
[b]*School of Physics and Technology, and Institute for Advanced Studies, Wuhan University, Wuhan 430072, China*
zpli@cnu.edu.cn; hxxu@whu.edu.cn

4.1 Introduction

Antennas were first developed a century ago as electrical devices to convert electric power into radio waves and vice versa. They play an essential role in modern wireless communication for radiating and receiving radio waves [1–4]. Plasmonic antennas as an analogue at the nanoscale are of great interest due to the unique ability of metallic nanostructures to manipulate the absorption and emission at optical frequency [4, 5], such as focusing light to volumes far beyond the diffraction limit [6–8], enhancing the Raman scattering and fluorescence of molecules [9, 10] and quantum emitters [11, 12], manipulating their emission polarization [10, 13, 14], and modifying

---

Nanophotonics: Manipulating Light with Plasmons
Edited by Hongxing Xu
Copyright © 2018 Pan Stanford Publishing Pte. Ltd.
ISBN 978-981-4774-14-7 (Hardcover), 978-1-315-19661-9 (eBook)
www.panstanford.com

their lifetime [15, 16]. Mediated by the surface plasmon resonance (SPR) of metallic nanostructures [17, 18], the plasmonic antennas can effectively convert the free propagating light into the nanoscale-enhanced near field [18–21] and vice versa, and a localized excitation can be coupled out as far-field radiation. The efficiency of a plasmonic antenna depends on its shape, material, dimension, assembly, and operation frequency [22, 23]. So far, various plasmonic antennas have been developed experimentally and theoretically to tune the SPR, such as nanodisks [24], triangles [25, 26], and flowers [27], as well as surface-plasmon-coupled antennas, including dimers [28], bowties [8], and trimers [10, 29–31]. These single and coupled plasmonic antennas have been investigated thoroughly by various far-field and near-field techniques, such as far-field scattering spectroscopy [22, 32, 33], electron energy loss spectroscopy [34, 35], nonlinear two-photon excited luminescence [9, 36], and near-field scanning optical microscopy. Abundant applications have been found in surface-enhanced Raman scattering (SERS) [6, 37], optical manipulation [28, 38], biosensing [39, 40], and integrated photonic devices [41–43].

In Section 4.2, we will first introduce single nanoantennas of various shapes and their far- and near-field properties. Then, in Section 4.3, the coupled antenna systems are considered, for example, coupled nanorods, bowties, and nanoaggregates. Nanoantennas of different geometries that can effectively control various light properties such as the intensity, direction, and polarization of emission are discussed.

4.2 Single Plasmonic Antennas

SPR is a kind of collective oscillation of a free electron cloud on the surface of metallic nanostructures, induced by incident light. It can enlarge effectively the cross section of nanostructures interacting with light. The absorption and scattering spectra of silver nanostructures of different shapes but with similar dimensions (both the side of the cube and the diameter of the ring are 50 nm) are shown in Fig. 4.1a, which presents a prominent resonant feature within the visible range due to the excitation of dipolar SPR. The cubic one has a broad peak at around 430 nm, whereas for the ring case, the peak is shifted to 568 nm. It demonstrates that the shape can greatly affect the frequency of SPR excited in metallic nanoparticles [5], which makes them good candidates for nanoantennas at different optical frequencies. Experimentally, the dependence of SPR on shape can be demonstrated by performing the single-particle spectroscopy. Figure 4.1b shows the continuous tunability of SPR from visible to infrared regions in anisotropic nanostructures by varying the aspect ratio [22]. Here, the widths of the particles are kept at around 55 nm, with the lengths varying from 100 to 200 nm. Two prominent resonance peaks can be observed in each single-particle dark-field scattering spectrum. One is located at around 430 nm and shows little dependence on the elongation of the length. This peak corresponds to the surface plasmon (SP) mode with charge oscillation along the short axis of the particle, that is, the transverse dipole plasmon mode. The other one red-shifts from 620 to 850 nm with the increase in the aspect ratio of particles. This peak is originated from the SP oscillation along the long axis of the particle, that is, the longitudinal dipole plasmon mode. Hence, it presents a great dependence on the particle length. Not only the tunable far-field radiation but also a huge local field enhancement can be obtained in the vicinity of plasmonic antennas under resonant excitation. In Fig. 4.1c, the calculated near field around a gold nanobar shows the excitation of the first three longitudinal plasmon modes supported by a 400 nm long silver nanobar with a 50 × 100 nm² cross section under the free-space excitation wavelengths λ_{free} = 1375, 770, and 630 nm. Here, the longitudinal resonances of a nanobar can be understood as an oscillation in the Fabry–Pérot cavity. The approximate relationship between the rod length L and the effective resonance wavelength λ_{eff} of longitudinal modes can be written as

$$L = \frac{n}{2}\lambda_{eff}, \quad (4.1)$$

where n is the integer number indicating the order of longitudinal plasmon resonance. The instantaneous charge configurations for the first three orders of longitudinal plasmon resonances are shown at the bottom of Fig. 4.1c.

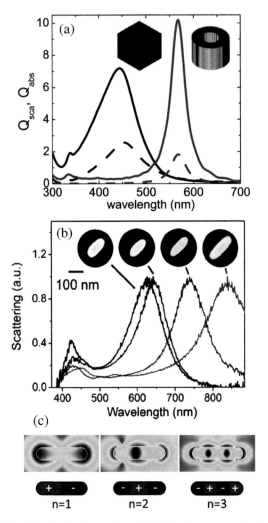

Figure 4.1 (a) Calculated absorption (solid lines) and scattering (dashed lines) efficiencies for a cubic-shaped and a ring-shaped silver nanoparticle. (b) Experimentally measured dark-field scattering spectra of single silver nanobars with different aspect ratios. (c) Electric field intensity enhancement corresponding to the first three orders of longitudinal plasmon resonances. The colors represent the field amplitudes on a logarithmic scale from 0.1 to 100. The bottom schematics show the instantaneous charge configurations for the first three longitudinal plasmon resonances. Reprinted with permission from Refs. [5, 22]. Copyright (2011, 2007) American Chemical Society.

The large near-field enhancement determines the ability of antennas to focus the energy into a subdiffraction-limited volume. Experimentally describing the near-field distribution is essential for the research and practical applications of plasmonic antennas. Nonlinear microscopy is a promising technique to visualize the highly confined electric field, because the higher power dependence of nonlinear responses on the electric field makes it preferentially sensitive to the most intense field close to the metal. For instance, the two-photon-induced luminescence (TPL) through interband transition of gold can provide an effective method to characterize the near-field distribution. The top panel of Fig. 4.2a shows an example of a TPL image from a single gold antenna under longitudinal excitation. Excited by the intense pulsed radiation at 730 nm, not only the longitudinal plasmon resonances, but also a TPL signal can be observed. The distribution of the TPL intensity around the antenna can be further mapped by scanning the antenna with respect to the illumination spot. Interestingly, it is found that the TPL signal is concentrated at the extremities of the gold bar antenna. The scanning electron microscopy (SEM) image of this nanobar is shown in the bottom panel of Fig. 4.2a. The superimposed black lines plot the TPL signal along the symmetry axis of the antennas. Owing to the quadratic dependence of the TPL signal on excitation intensity I^2, the observed TPL map should well reflect the distribution of the local field to the fourth power $|E|^4$. The top panel of Figure 4.2b shows the calculated distribution of $|E|^4$ for resonant plasmon modes. Additionally, it is known that the optical resolution in experiments is about 200 nm. So the calculated electric field map is further convoluted by a 2D Gaussian profile with a 200 nm full-width at half-maximum (FWHM). The resulting field distribution is fully dominated by the strong intensity at the extremities, as shown in the bottom panel of Fig. 4.2b. By comparison, it is found that the TPL image can well reflect the local field distributions of resonant plasmon modes, which provides a useful technique for the highly confined electric field visualization around the gold nanoantennas [9].

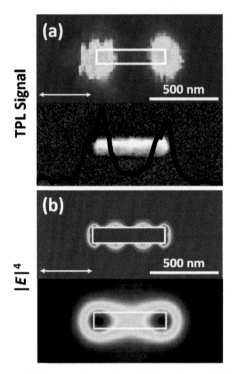

Figure 4.2 (a, top) TPL scans for a single 500 nm long gold bar under pulsed laser excitation at 730 nm. (a, bottom) the SEM image. (b, top) Calculated $|E|^4$ distribution for resonant plasmon modes under longitudinal polarization at 710 nm excitation; (b, bottom) convoluted $|E|^4$ distribution by a 2D Gaussian profile with 200 nm FWHM. The arrows indicate the incident polarization. Reprinted with permission from Ref. [9]. Copyright (2008) by the American Physical Society.

Besides the TPL image, the near-field distribution can also be visualized by the scanning near-field optical microscope (SNOM). As mentioned before, the shape of the antenna influences both far-field and near-field properties [25]. Figure 4.3a shows the dark-field scattering spectrum of a triangular nanoantenna with edge length 200 nm. Three typical SEM images of triangle antennas of that size are shown in the insets. The scale bar is 200 nm. Two peaks dominate the spectrum. One is at ~670 nm and the other at ~430 nm, corresponding to the dipole resonance and quadrupole dark mode, respectively. Here, we should note that the quadrupole mode becomes radiative through hybridization with the dipole mode. Figure 4.3b shows the size effect on the resonant frequencies of dipolar and quadrupolar plasmon modes. The dipolar and quadrupolar resonances progressively shift to lower energies with the increase of the edge length.

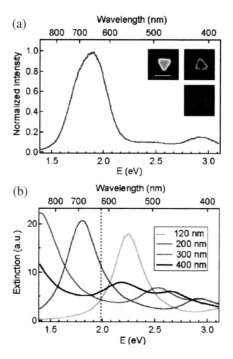

Figure 4.3 (a) The normalized scattering spectrum of an isolated silver nanotriangle with edge length 200 nm. (b) The scattering spectra of nanotriangles with edge lengths 120, 200, 300, and 400 nm. Reprinted with permission from Ref. [25]. Copyright (2008) American Chemical Society.

The size dependence of these two modes is then visualized by an SNOM under an excitation of 633 nm, indicated by the red dashed line in Fig. 4.3b. Figure 4.4a shows the SNOM image of a silver nanotriangle with edge length ~120 nm. The atomic force microscopy (AFM) topography is shown in the inset. Under an s-polarized incidence of 633 nm, the in-plane dipolar resonance can be excited, as discussed in Fig. 4.3b. The observed near-field pattern exhibits prominent dipolar resonance features, with the strongest local field at two tips and extended spatial distribution beyond the particle boundaries, which is consistent with the field distribution of a typical dipole. The numerical simulation in Fig. 4.4b using the discrete dipole approximation (DDA) also gives the dipolar resonance features (the plus and minus signs indicate their relative phases), which corresponds well to the SNOM results. For the triangle with a larger edge length ~450 nm, the resonance is dominated by the quadrupole plasmon mode under the 633 nm excitation, as indicated in Fig. 4.3b. Figure 4.4c shows the corresponding SNOM result, and the AFM topography image is shown in the inset. Under the s-polarized incidence, the SNOM image shows a typical quadrupole response featured by four local-field-enhanced regions. It gives rise to the largest near-field enhancement at the lower two adjacent edges. And the other two weaker charge centers of the quadrupole can be identified at the two tips of the top edge. The DDA simulation under the experimental conditions also shows good accordance with the experiments, as shown in Fig. 4.4d. These results demonstrate that an SNOM can also well visualize the local electric field distributions of resonant plasmon modes under a certain excitation.

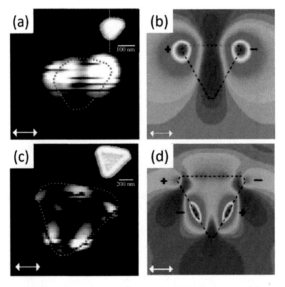

Figure 4.4 (a, b) Experimental and theoretical distributions of the local electric field for a nanotriangle (edge length ~120 nm) supporting a dipole plasmon resonance. Panels (c) and (d) are identical to panels (a) and (b) but for a larger nanotriangle (edge length ~450 nm) under quadrupole excitation. Reprinted with permission from Ref. [25]. Copyright (2008) American Chemical Society.

4.3 Coupled Optical Antennas

4.3.1 Control of Local Intensity

Optical near-field coupling between closely spaced plasmonic metal nanostructures is important to a range of nanophotonic applications, including surface-enhanced molecular spectroscopy [6], nano-optical sensing [44], and various novel light-harvesting concepts [45]. The coupled plasmonic metallic nanostructures can concentrate light in well-defined electromagnetic hot spots, which is a common way to enhance the performance of nanoantennas. Figure 4.5 shows two 500 nm long gold rods with a 40 nm gap. The coupling between the two rods gives rise to strong local enhancement, which is shown by the intensity of TPL in Fig. 4.5a [9]. The SEM images of this structure are shown in Fig. 4.5b. The superimposed black curves plot the TPL signal along the symmetry axis of the rods' antennas. More experiments and theoretical calculations indicate that the intensity enhancement at the visible frequencies can be as high as 10^3 when the gap is reduced to a few of nanometers. For these kinds of rods' dimer antennas, designed to be resonant at optical frequencies [2], strong field enhancement in the gap could also lead to white-light supercontinuum generation. The antenna length at resonance is shorter than half of the wavelength of the incident light.

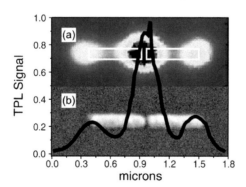

Figure 4.5 (a) TPL map for two 500 nm long gold rods with a 40 nm gap and (b) SEM image. Reprinted with permission from Ref. [9]. Copyright (2008) by the American Physical Society.

The bowtie structure [15] shown in the SEM image in Fig. 4.6a is one of the most investigated configurations in coupled nanoantennas. According to the electromagnetic calculation, these kinds of metallic bowtie nanoantennas should provide optical fields that are confined to spatial scales far below the diffraction limit, as shown in Fig. 4.6b. The enhancement factor of $|E|^2/|E_0|^2$ at its gap is evaluated by the surface-enhanced fluorescence, which is shown in Fig. 4.6c. A schematic representation of molecules randomly placed around a gold bowtie nanoantenna on a transparent substrate is shown in the inset. The size of the gap varies from 15 nm to 80 nm. The maximum enhancement of fluorescence can reach 1340-fold. Also the Au bowties lithographically fabricated with lengths and gaps of both tens of nanometers can improve the mismatch between optical wavelengths and nanoscale objects [36]. Enhancements larger than 10^3 occur for a 20 nm gap when the length of the bowties is 75 nm, which is in good agreement with theoretical simulations.

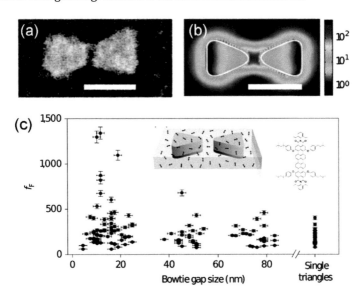

Figure 4.6 (a) SEM image of a gold bowtie nanoantenna (scale bar = 100 nm). (b) Calculated distribution of the local electric field intensity enhancement. (c) Enhancement factor as a function of the gap size. Reprinted by permission from Macmillan Publishers Ltd: [*Nature Photonics*] (Ref. [15]), copyright (2009).

A more systematic study was done by Schnell et al., as shown in Fig. 4.7, where they investigated the near-field oscillations of single to coupled plasmonic antennas at infrared frequency by a scattering-type scanning near-field optical microscope (s-SNOM) [4], which provides information on both the amplitude and the phase of local fields in samples of interest [46]. The near-field images show the z component of the electric field in amplitude E_z and phase φ_z in a height of 50 nm above the rod surface. In Fig. 4.7a, the dipolar mode excited at a nanorod has been clearly revealed by amplitude signals at both ends of the nanorod and phase change at the rod center. The near-field patterns of the nanorod are completely changed when a wedge is cut in the middle of the nanorod, where a thick metal bridge connects the two antenna segments, that is, a low-impedance loaded antenna. No significant near fields near the gap are observed, and the dipolar mode of the continuous nanorod still holds, which is shown in Fig. 4.7b. This might be caused by a current flow through the bridge that prevents charge at the gap. But if the rod is cut more deeply or even fully cut to form a high-impedance loaded antenna, where a tiny metal bridge connects the two antenna segments, the case is completely different. The nanorod gradually changes to a coupled nanoantenna from a single nanoantenna. In Fig. 4.7c, the bridge between the two parts of the nanorods is only 2% of the cross section, which cannot maintain the dipolar mode any more. The near-field image is different from both the continuous and fully cut rods, which shows a significant amplitude along each segment and a phase gradient of about 80°. At last, for the fully cut rod in Fig. 4.7d, an 80 nm gap is formed in the rod center. Each nanoantenna segment oscillates as a dipole. Hence, a face change appears between the two segments and the gap.

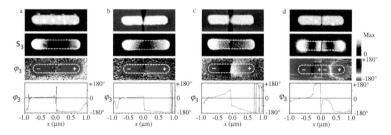

Figure 4.7 Near-field images of progressively loaded nanoantennas at a wavelength of λ = 9.6 µm. (a) Continuous-rod antenna. (b) Low-impedance loaded antenna. (c) High-impedance loaded antenna. (d) Fully cut antenna where the two antenna segments are completely separated. The images show experimental results of topography and near-field amplitude s_3 and phase φ_3. Reprinted by permission from Macmillan Publishers Ltd: [*Nature Photonics*] (Ref. [4]), copyright (2009).

Electromagnetic field localization and enhancement in coupled nanoantennas can also result in strong nonlinear responses [47] that depend strongly on the antenna shape, size, and gap. These distinct properties are a consequence of the large electromagnetic field concentration at the SPR frequency, which can enhance the local strength of light–matter interactions. We have shown that TPL can be used to image the near-field distribution around metallic nanostructures. Here we focus more on the nonlinear effects. The second harmonic generation (SHG) from plasmonic dimer nanoantennas is extremely sensitive to asymmetry in the nanostructure shape, even if its linear response is barely modified [48]. Minute geometry asymmetry and surface roughness are revealed by far-field analysis, demonstrating that the SHG is a promising tool for the sensitive optical characterization of plasmonic nanoantennas. For a large range of geometries, the large third harmonic generation (THG) [49] in plasmonic dimer nanoantennas is determined by the linear response. An excellent agreement is found between the measured spectra and a simple nonlinear oscillator model. Deviations from the model occur for gap sizes below 20 nm, indicating that only for small distances the hot spots contribute noticeably to the THG. The emission of the SHG could also be remotely excited in micrometer-long or coupled gold rod antennas by a tightly focused near-infrared femtosecond laser [50]. The nonlinear responses are locally induced by the propagating SPs at the excitation frequency, enabling the tailoring of nonlinear responses.

4.3.2 Control of Emission Direction

One of the most important properties of nanoantennas is controlling the far-field emission from both an emitter and nanostructures. In Fig. 4.8, a Yagi–Uda plasmonic antenna can beam the emission of a quantum dot positioned near one of the arms of the antenna [11]. Figure 4.8a shows the SEM images of the nanoantenna composed by five gold nanobars. The distance between each arm in a Yagi–Uda antenna is about 100 nm. A quantum dot is attached to one end of an arm, indicated with a square. The emission direction involves a large solid angle for the quantum dot emitter. Figure 4.8b shows the radiation intensity distribution at the back focal plane of the objective after an 830 nm long-pass filter. The angular distribution is shown in Fig. 4.8c, which is consistent well with the simulation result. The constructive interference of the emissions from each arm excited by the emitter results in a narrow directional radiation pattern.

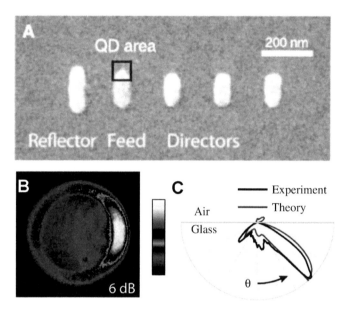

Figure 4.8 (a) SEM image of a five-element Yagi–Uda gold nanoantenna. (b) Radiation pattern from a Yagi–Uda antenna. (c) Measurement and calculation results of the angular radiation distribution for the antenna. From Ref. [11]. Reprinted with permission from AAAS.

In the Yagi–Uda gold nanoantenna, elements are well separated from each other, which is relatively large for plasmonic coupling. The emission direction could also be well controlled by stronger coupling. In Fig. 4.9 [51], the scattering properties can be tuned by two closely spaced metal disks of different materials with appropriate wavelength-dependent permittivities, for example, silver and gold, but the effect is not limited to these materials. The concept is also restricted neither to disk-shaped nanoparticles nor to planar dimers. This bimetallic dimer shows a novel degree of freedom for manipulating the phase or additive phase accumulation by introducing an asymmetric material composition and thus material-dependent SPR. In spite of being as compact as $\sim\lambda^3/100$, this bimetallic nanoantenna shows exotic optical properties, in particular wavelength-dependent directional scattering. Red and blue light will be irradiated in opposite directions, that is, color routing. In the inset of Fig. 4.9b, corresponding Fourier color images are shown at two specific wavelengths, 450 nm and 700 nm. Interestingly, the direction of the scattering is dependent on the wavelength. For blue (450 nm) and red light (700 nm), the scattering is in opposite directions and the switching point between scattering to the right and to the left appears at around 500 nm. These effects are similar with the incident polarization parallel or perpendicular to the dimer axis. These bimetallic nanoantennas fabricated through cheap colloidal lithography offer a versatile platform for manipulating optical response through wavelength, materials, and geometry, which can be used as ultrasmall ($\sim\lambda^3/100$) photon couplers and sorters, directional emitters, sensors, and metamaterials.

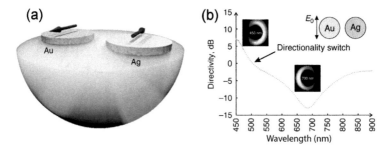

Figure 4.9 (a) An artist's view of a bimetallic nanodisk dimer. (b) Experimental scattering directivities as a function of the wavelength. Reprinted by permission from Macmillan Publishers Ltd: [*Nature Communications*] (Ref. [51]), copyright (2011).

4.3.3 Control of Far-Field Polarization

Not only the near-/far-field intensity and emission direction but also the light polarization from an emitter (e.g., Raman scattering [RS] of molecules) in the nanogap can be manipulated significantly [10, 13]. Firstly, the simplest case of a nanocrystal dimer is considered. Figure 4.10a shows the SEM image of such a dimer. The angle of the dimer axis is ~130° with respect to the x axis, which corresponds to the maximal RS intensity shown in the Fig. 4.10b. It demonstrates the favorite incident polarization for the enhancement is along the dimer axis for the Raman shift (773 cm^{-1} and 1650 cm^{-1}). Here the low concentration of molecules used ensures that each nanoantenna aggregate contains no more than a single molecule [52]. The depolarization ρ in Fig. 4.10c is defined as $\rho = (I_\parallel + I_\perp)/(I_\parallel + I_\perp)$, where $I_\parallel$ and $I_\perp$ are RS signals with orthogonal polarization, which indicates that the polarization of the RS light is also along the dimer axis at these two wavelengths (555 nm and 583 nm). No wavelength dependence is observed, and the maximal depolarization ratio is polarized along the dimer axis. These properties of a dimer can be simulated by generalized Mie theory through treating the nanoparticles as spheres. As the single-molecule Raman signal can only come from hot spots, the calculations mainly include the field enhancement in the junction of the dimer and the depolarization of the emission from a dipole emitter situated in this junction. The calculated results shown by the curves in Figs. 4.10b and 4.10c are consistent with the experiment. This far-field polarization is expected due to the axial symmetric geometry of the dimer and had been mentioned in many previous reports [53–55], which also justifies our results.

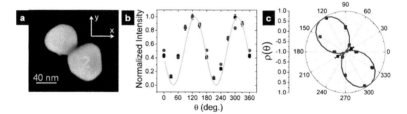

Figure 4.10 Polarization response of a nanocrystal dimer. (a) SEM image showing a dimer of nanoparticles. (b) Normalized SERS intensity at 555 nm (squares) and 583 nm (circles) as a function of the angle of rotation by the $\lambda/2$-wave plate. (c) Depolarization ratio ρ measured at 555 nm (squares) and 583 nm (circles). Reproduced from Ref. [10]. Copyright (2008) National Academy of Sciences.

As to the nanocrystal trimer, a drastically different behavior is shown in Fig. 4.11. The intensity in Fig. 4.11b is maximal at the angle of ~75°, which is close to the axis of nanoparticles 1 and 2. The depolarization ratio profiles of two wavelengths both rotate with respect to the intensity profile, but they are not coincident. Depolarization profiles are wavelength dependent in this case and are aligned differently than the intensity profiles. The corresponding simulation shows that only when the molecule is in the junction between particles 1 and 2, pointed with a red arrow in Fig. 4.11a, the calculated results are in good agreement with the experiments for both normalized intensity and depolarization, which also confirms the assumption that only one molecule in the junction contributes to the RS signal. This counterintuitive wavelength-dependent polarization rotation is not an accident, which only exists in the cases when the number of metal particles in the nanoantenna is larger than two.

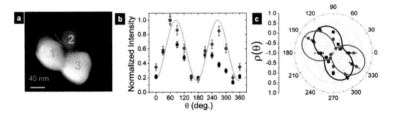

Figure 4.11 Polarization response of a nanocrystal trimer. (a) SEM image of the trimer. (b) Normalized SERS intensity at 555 nm (squares) and 583 nm (circles) as a function of the angle of rotation by the $\lambda/2$-wave plate. (c) Depolarization ratio ρ measured at 555 nm (squares) and 583 nm (circles). Reproduced from Ref. [10]. Copyright (2008) National Academy of Sciences.

Different from the symmetric dimer case, the third nanoparticle in the trimer breaks the axial symmetry of the dimer. For the SPs excited by a laser, the electric field distribution for a trimer system under different incident polarization is shown in Fig. 4.12. The field enhancement in the junction of the trimer is not so sensitive to the incident polarization as the dimer case. Although the enhancement in the junction of particles 2 and 3 is reduced due to the existence of the relatively large particle 1, the local electric field in this junction is still along the dimer axis, irrespective of the incident polarization [56, 57].

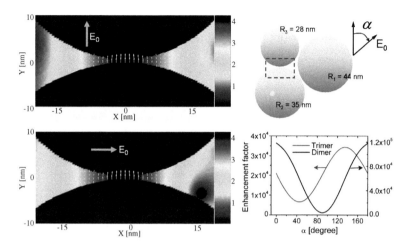

Figure 4.12 The local field in one of the junctions of a silver trimer under excitations with different polarization (laser wavelength = 532 nm). Reproduced from Ref. [10]. Copyright (2008) National Academy of Sciences.

For the SPs excited by the molecule dipole, on the other hand, the influence of the third nanoparticle to the nanoantenna system is demonstrated in Fig. 4.13a. The depolarization of the emission from the dipole in the gap is calculated for different distances between the third nanosphere and the dimer. When the third sphere is far away from the dimer, the depolarization of emission is just like the dimer case. However, as the third sphere moves closer to the dimer, the polarization of the emission is rotated effectively with respect to the original dimer axis. This degree of the rotation is obviously wavelength dependent. This kind of rotation of polarization is caused by the increasing SP coupling between the third sphere and the dimer. Even more interesting, an overall rotation (about 2π) could be achieved by tuning the distance of the third nanoparticle to the original dimer. Moreover, at a very small distance (e.g., $d = 0.5$ nm), the emitted light is almost circularly polarized.

On the basis of these simulations, it can be inferred that laser-induced SPs in the trimer could create quasi-isotropic hot spots [56, 57] and the molecule-dipole-induced SPs of the trimer may cause the rotation of the polarization of RS light. Here, the asymmetric trimer actually acts as a birefringence crystal worked at the nanometer scale. The effect of polarization rotation can be used as an equivalent half-wave plate for the RS emission.

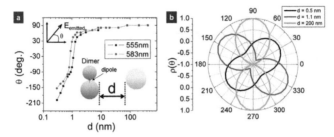

Figure 4.13 Polarization rotation as a function of the position of the third particle in a trimer. (a) Polarization rotation as the right particle in the inset diagram approaches the other two particles. (b) Depolarization ratio patterns for the case of d = 200 nm, 1.1 nm, and 0.5 nm. Reproduced from Ref. [10]. Copyright (2008) National Academy of Sciences.

Besides the distance, the size of the third nanosphere could also affect the SPs excited in the nanoantenna. As shown in Fig. 4.14, a third sphere with a fixed distance to the former dimer varies its radius from 1 nm to 100 nm. When the size of the third sphere is small, its influence on the SP response of the dimer is very weak; hence in this case the depolarization of the dipole emission is still along the dimer axis. On the other hand, when the size of the third sphere is big, the three spheres show strong coupling, which induces a wavelength-dependent polarization rotation, and an abrupt change occurs at a specific radius range.

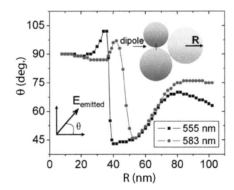

Figure 4.14 Polarization rotation as a function of the size of the third particle in a trimer, while keeping all interparticle distances constant at d = 1 nm. Reproduced from Ref. [10]. Copyright (2008) National Academy of Sciences.

Here, the orientation of the molecule on the metal surface is also considered. Figure 4.15 shows the far-field intensity enhancement as the rotation of the molecule relative to the dimer axis. It demonstrates that only the dipole component parallel to the dimer axis could be amplified significantly by the aggregates' nanoantenna. That is, the far-field polarization of the dipole emission from the nanoantenna is independent of the assumed orientation of the dipole. Hence, it is impossible that the wavelength-dependent rotation of the depolarization of the RS light presented originated from the orientation of the molecule or its polarizability tensor [57].

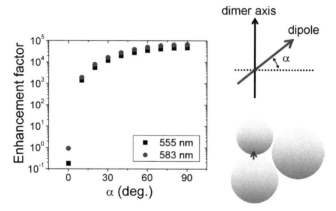

Figure 4.15 The intensity of the enhanced field emitted by an oscillating dipole, as detected in the far field. Reproduced from Ref. [10]. Copyright (2008) National Academy of Sciences.

Besides planar metal nanoparticles, there are variable kinds of nanostructures to tune the far-field polarization. For example, in Fig. 4.16, monolayers of 3D chiral heterotrimers and heterotetramers composed of coupled metal nanodisks of different heights fabricated by facile hole-mask colloidal lithography show hot spots, circular dichroism (CD), and optical chirality [58]. The extinction cross-sectional spectra (for right-circularly polarized [RCP] and left-circularly polarized [LCP] light) of the right-handed (RH) and left-handed (LH) silver tetramers are displayed in Fig. 4.16a and 4.16b, respectively. The dissymmetry factor $g = \Delta T/\overline{T}$ shows the relative extinction differences. In Fig. 4.16c, the CD spectra are reverse, that is, the extinction of the RH tetramer excited by RCP/LCP light is similar to that of the LH tetramer excited by LCP/RCP light. Its CD is due to strong near-field coupling and intricate phase retardation effects.

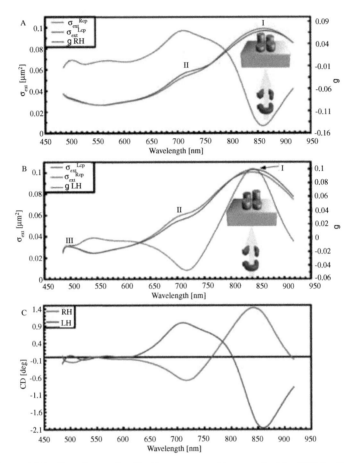

Figure 4.16 Experimental extinction cross sections excited by RCP (red) and LCP (blue) light and dissymmetry factors g (gray) for (A) RH silver tetramers and (B) LH tetramers and (C) CD spectra for the RH (red) and LH (blue) structures. Reprinted with permission from Ref. [58]. Copyright (2014) American Chemical Society.

4.4 Summary

Nanoscale antennas are optical devices that can facilitate the localization and transfer of electromagnetic energy at the nanometer length scale. Typical single and coupled nanoantennas have been introduced here. Plasmonic antennas can be engineered to manipulate light properties at the nanoscale. For single nanoantennas, the far-field spectrum is dependent on size, shape, materials, etc. Dipolar and higher modes can be excited in metallic nanostructures, which results in a corresponding near-field pattern that can by observed by TPL and scanning probe techniques. More importantly, nanoantennas are often based on coupled structures, such as nanorod dimers with a small gap, bowties, Yagi–Uda structures, and nanoparticle aggregates. With the help of SP coupling, light intensity, direction, and polarization can be well tailored. We can expect that optical antennas link propagating radiation and local optical fields; they should find applications in optical characterization, manipulation of nanostructures, and optical information processing.

References

1. J. J. Greffet, *Science*, **308**(5728), 1561 (2005).
2. P. Mühlschlegel, H.-J. Eisler, O. J. F. Martin, B. Hecht, and D.W. Pohl, *Science*, **308**(5728), 1607 (2005).
3. T. Shegai, V. D. Miljković, K. Bao, H. X. Xu, P. Nordlander, P. Johansson, and M. Käll, *Nano Lett.*, **11**(2), 706 (2011).
4. M. Schnell, A. García-Etxarri, A. J. Huber, K. Crozier, J. Aizpurua, and R. Hillenbrand, *Nat. Photonics*, **3**(5), 287 (2009).
5. V. Giannini, A. I. Fernandez-Dominguez, S. C. Heck, and S. A. Maier, *Chem. Rev.*, **111**(6), 3888 (2011).
6. H. X. Xu, E. J. Bjerneld, M. Käll, and L. Börjesson, *Phys. Rev. Lett.*, **83**(21), 4357 (1999).
7. Z. Zhang, A. Weber-Bargioni, S. W. Wu, S. Dhuey, S. Cabrini, and P. J. Schuck, *Nano Lett.*, **9**(12), 4505 (2009).
8. D. P. Fromm, A. Sundaramurthy, P. J. Schuck, G. Kino, and W. E. Moerner, *Nano Lett.*, **4**(5), 957 (2004).
9. P. Ghenuche, S. Cherukulappurath, T. H. Taminiau, N. F. van Hulst, and R. Quidant, *Phys. Rev. Lett.*, **101**(11), 116805 (2008).
10. T. Shegai, Z. P. Li, T. Dadosh, Z. Y. Zhang, H. X. Xu, and G. Haran, *Proc. Natl. Acad. Sci. USA*, **105**(43), 16448 (2008).
11. A. G. Curto, G. Volpe, T. H. Taminiau, M. P. Kreuzer, R. Quidant, and N. F. van Hulst, *Science*, **329**(5994), 930 (2010).

12. M. Ringler, A. Schwemer, M. Wunderlich, A. Nichtl, K. Kürzinger, T. Klar, and J. Feldmann, *Phys. Rev. Lett.*, **100**(20), 203002 (2008).
13. Z. P. Li, T. Shegai, G. Haran, and H. X. Xu, *ACS Nano*, **3**(3), 637 (2009).
14. L. K. Yang, H. C. Wang, Y. Fang, and Z. P. Li, *ACS Nano*, **10**(1), 1580 (2016).
15. A. Kinkhabwala, Z. F. Yu, S. H. Fan, Y. Avlasevich, K. Mullen, and W. E. Moerner, *Nat. Photonics*, **3**(11), 654 (2009).
16. J. N. Farahani, D. W. Pohl, H. J. Eisler, and B. Hecht, *Phys. Rev. Lett.*, **95**(1), 017402 (2005).
17. K. G. Lee, H. W. Kihm, J. E. Kihm, W. J. Choi, H. Kim, C. Ropers, D. J. Park, Y. C. Yoon, S. B. Choi, D. H. Woo, J. Kim, B. Lee, Q. H. Park, C. Lienau, and D. S. Kim, *Nat. Photonics*, **1**(1), 53 (2007).
18. M. W. Knight, N. K. Grady, R. Bardhan, F. Hao, P. Nordlander, and N. J. Halas, *Nano Lett.*, **7**(8), 2346 (2007).
19. Z. P. Li, F. Hao, Y. Z. Huang, Y. R. Fang, P. Nordlander, and H. X. Xu, *Nano Lett.*, **9**(12), 4383 (2009).
20. Z. P. Li, S. P. Zhang, N. J. Halas, P. Nordlander, and H. X. Xu, *Small*, **7**(5), 593 (2011).
21. Z. P. Li, K. Bao, Y. R. Fang, Z. Q. Guan, N. J. Halas, P. Nordlander, and H. X. Xu, *Phys. Rev. B*, **82**(24), 241402 (2010).
22. B. J. Wiley, Y. C. Chen, J. M. McLellan, Y. J. Xiong, Z.-Y. Li, D. Ginger, and Y. N. Xia, *Nano Lett.*, **7**(4), 1032 (2007).
23. H. Y. Liang, H. X. Yang, W. Z. Wang, J. Q. Li, and H. X. Xu, *J. Am. Chem. Soc.*, **131**(17), 6068 (2009).
24. C. Langhammer, B. Kasemo, and I. Zoric, *J. Chem. Phys.*, **126**(19), 194702 (2007).
25. M. Rang, A. C. Jones, F. Zhou, Z.-Y. Li, B. J. Wiley, Y. N. Xia, and M. B. Raschke, *Nano Lett.*, **8**(10), 3357 (2008).
26. J. Nelayah, M. Kociak, O. Stéphan, F. J. García de Abajo, M. Tencé, L. Henrard, D. Taverna, I. Pastoriza-Santos, L. M. Liz-Marzán, and C. Colliex, *Nat. Phys.*, **3**(5), 348 (2007).
27. H. Y. Liang, Z. P. Li, W. Z. Wang, Y. S. Wu, and H. X. Xu, *Adv. Mater.*, **21**(45), 4614 (2009).
28. F. Svedberg, Z. P. Li, H. X. Xu, and M. Käll, *Nano Lett.*, **6**(12), 2639 (2006).
29. R. C. Jin, Y. W. Cao, C. A. Mirkin, K. L. Kelly, G. C. Schatz, and J. G. Zheng, *Science*, **294**(5548), 1901 (2001).
30. J. Aizpurua, P. Hanarp, D. Sutherland, M. Käll, G. Bryant, and F. García de Abajo, *Phys. Rev. Lett.*, **90**(5), 057401 (2003).

31. F. Hao, C. L. Nehl, J. H. Hafner, and P. Nordlander, *Nano Lett.*, **7**(3), 729 (2007).
32. L. J. Sherry, S. H. Chang, G. C. Schatz, R. P. Van Duyne, B. J. Wiley, and Y. N. Xia, *Nano Lett.*, **5**(10), 2034 (2005).
33. L. Chuntonov and G. Haran, *Nano Lett.*, **11**(6), 2440 (2011).
34. A. L. Koh, A. I. Fernandez-Dominguez, D. W. McComb, S. A. Maier, and J. K. Yang, *Nano Lett.*, **11**(3), 1323 (2011).
35. J. A. Scholl, A. L. Koh, and J. A. Dionne, *Nature*, **483**(7390), 421 (2012).
36. P. J. Schuck, D. P. Fromm, A. Sundaramurthy, G. S. Kino, and W. E. Moerner, *Phys. Rev. Lett.*, **94**(1), 017402 (2005).
37. W. Wang, Z. P. Li, B. H. Gu, Z. Y. Zhang, and H. X. Xu, *ACS Nano*, **3**(11), 3493 (2009).
38. Z. P. Li, M. Käll, and H. X. Xu, *Phys. Rev. B*, **77**(8), 085412 (2008).
39. S. Lal, S. E. Clare, and N. J. Halas, *Acc. Chem. Res.*, **41**(12), 1842 (2008).
40. C. Loo, A. Lowery, N. J. Halas, J. West, and R. Drezek, *Nano Lett.*, **5**(4), 709 (2005).
41. E. Ozbay, *Science*, **311**(5758), 189 (2006).
42. F. J. Garcia de Abajo, J. Cordon, M. Corso, F. Schiller, and J. E. Ortega, *Nanoscale*, **2**(5), 717 (2010).
43. R. Kirchain and L. Kimerling, *Nat. Photonics*, **1**(6), 303 (2007).
44. F. Hao, Y. Sonnefraud, P. V. Dorpe, S. A. Maier, N. J. Halas, and P. Nordlander, *Nano Lett.*, **8**(11), 3983 (2008).
45. A. Aubry, D. Y. Lei, A. I. Fernandez-Dominguez, Y. Sonnefraud, S. A. Maier, and J. B. Pendry, *Nano Lett.*, **10**(7), 2574 (2010).
46. T. Neuman, P. Alonso-González, A. Garcia-Etxarri, M. Schnell, R. Hillenbrand, and J. Aizpurua, *Laser Photonics Rev.*, **9**(6), 637 (2015).
47. J. Y. Suh and T. W. Odom, *Nano Today*, **8**(5), 469 (2013).
48. J. Butet, K. Thyagarajan, and O. J. Martin, *Nano Lett.*, **13**(4), 1787 (2013).
49. M. Hentschel, T. Utikal, H. Giessen, and M. Lippitz, *Nano Lett.*, **12**(7), 3778 (2012).
50. S. Viarbitskaya, O. Demichel, B. Cluzel, G. Colas des Francs, and A. Bouhelier, *Phys. Rev. Lett.*, **115**(19), 197401 (2015).
51. T. Shegai, S. Chen, V. D. Miljković, G. Zengin, P. Johansson, and M. Käll, *Nat. Commun.*, **2**, 481 (2011).
52. E. C. Le Ru, M. Meyer, and P. G. Etchegoin, *J. Phys. Chem. B*, **110**, 1944 (2006).

53. T. Itoh, K. Hashimoto, and Y. Ozaki, *Appl. Phys. Lett.*, **83**(11), 2274 (2003).
54. A. M. Michaels, J. Jiang, and L. Brus, *J. Phys. Chem. B*, **104**(50), 11965 (2000).
55. J. Jiang, K. Bosnick, M. Maillard, and L. Brus, *J. Phys. Chem. B*, **107**(37), 9964 (2003).
56. H. X. Xu, *J. Opt. Soc. Am. A*, **21**(5), 804 (2004).
57. P. G. Etchegoin, C. Galloway, and E. C. Le Ru, *Phys. Chem. Chem. Phys.*, **8**(22), 2624 (2006).
58. R. Ogier, Y. R. Fang, M. Svedendahl, P. Johansson, and M. Käll, *ACS Photonics*, **1**(10), 1074 (2014).

Chapter 5

Plasmon-Assisted Optical Forces

Lianming Tong[a] and Hongxing Xu[b]

[a]*Center for Nanochemistry, College of Chemistry and Molecular Engineering, Peking University, Beijing 100871, China*
[b]*School of Physics and Technology, and Institute for Advanced Studies, Wuhan University, Wuhan 430072, China*
tonglm@pku.edu.cn; hxxu@whu.edu.cn

5.1 Introduction

Optical force has opened up the unique possibility of manipulating objects of micro-/nanoscale sizes by means of light [1–6]. In general, the forces include a gradient type, which originates from the electric field gradient of light, and a dissipative type, also known as the radiation pressure, a result of photons impinging on and absorbed by the objects. A typical application of optical forces is the so-called optical tweezer that was discovered by Ashkin [7, 8] in 1971 and have nowadays become a powerful tool in optical trapping and manipulation, for example, sorting biological cells, measuring biophysical properties of single biomacromolecules,

---

Nanophotonics: Manipulating Light with Plasmons
Edited by Hongxing Xu
Copyright © 2018 Pan Stanford Publishing Pte. Ltd.
ISBN 978-981-4774-14-7 (Hardcover), 978-1-315-19661-9 (eBook)
www.panstanford.com

and manipulating dielectric and metallic nanoparticles [4, 9–15]. In nanostructures that support surface plasmons, an enhanced electromagnetic (EM) near field is induced under resonant excitation, leading to a series of interesting applications in surface-enhanced spectroscopy and sensing [16–21]. The induced EM fields are closely bound to the surface and attenuate dramatically with increasing distance from the surface, thus forming a sharp gradient that gives rise to an attractive optical force proportional to the gradient [3, 22, 23]. For propagating surface plasmons, the scattering force is also an interesting component to study—it pushes the trapped particle to move along the propagation direction, for example, on the surface of metal thin films and nanostripes [24, 25].

Plasmon-assisted optical forces can be discussed in two slightly different categories: with and without a focused laser beam that provides additional gradient force from the laser itself. The former mainly refers to laser trapping of metal nanoparticles, where surface plasmons take effects in such a way that the behavior of the trapped particles differs substantially compared to dielectric ones and the interaction between the trapped metal nanoparticles involves the optical component resulting from the near-field coupling. The latter deals with optical trapping and manipulation using metal nanostructures fabricated on a substrate. In this case, an unfocused or slightly focused laser beam is used only to excite surface plasmons in the structures but not to provide any additional trapping effect.

There have been a number of publications on both theoretical and experimental work in the field of surface-plasmon-assisted forces, for example, optical forces of spherical metal nanoparticles in an optical trap of laser tweezers [10, 11, 14, 26, 27], optical trapping of dielectric/metal nanoparticles using metal nanostructures or metal thin films [3, 28–30], and plasmon-driven nanomotors [31]. In this chapter, the plasmon-assisted optical forces will be discussed in the following content: in Section 5.2, we will discuss the theoretical background of optical forces between metal nanoparticles due to near-field coupling; Section 5.3 gives an overview of the recent progress in experimental demonstrations of surface-plasmon-assisted optical forces; and the summary and outlook are given in Section 5.4.

5.2 Theoretical Calculations on Optical Forces in Near-Field-Coupled Nanoparticles

Great efforts have been made in the theoretical calculation of optical forces of metal nanostructures, for example, spherical and bipyramidal metal nanoparticle dimers [32–35] and aggregates of metal nanoparticles [36], single and dimers of nanodisks [37], and metal thin films [28, 38]. Typically, the EM field is first calculated using electrodynamic methods such as generalized Mie theory (GMT), finite-difference time-domain (FDTD), or discrete dipole approximation (DDA). The force can then be obtained using Maxwell's stress tensor formalism or the Lorentz force method. To define the stability of an optical trap, the optical potential can be calculated by integrating the optical force from the point of interest to infinitive and is usually given in the unit of thermal energy at room temperature k_BT, where k_B is the Boltzmann constant and T is the room temperature.

In 2002, Xu et al. [22] showed the theoretical calculations of optical potential and forces between spherical metal nanoparticles. In Fig. 5.1a, the spatial variation of optical potential U/k_BT around a chain of three nanoparticles in water is demonstrated. The radius of Ag nanoparticles is 25 nm, and the gap distances are both 1 nm. It was excited at resonance by a 760 nm laser at 10 mW/µm^2 power density. It is seen that the strength of the optical potential is the strongest at the nanogap where the EM field is the strongest. Accordingly, it can be anticipated that a single molecule that resides within or passes through the trapping volume would be instantaneous trapped at the gaps. The optical force at the gap of a dimer of Ag nanoparticles (schemed in the inset: radius 45 nm and gap distance 1 nm) under resonance illumination (wavelength 550 nm) is shown in Fig. 5.1b. F_C and F_D represent the gradient and dissipative forces, respectively. The direction of the forces can also be found, that is, the gradient force always points to the gap center (positive to the left and negative to the right), while the scattering force is always along the incident **k**-vector (positive values). On the other hand, the gradient force is more than 1 order of magnitude stronger than the scattering force. These results provide theoretical support for single-molecule surface-enhanced Raman scattering (SERS) in the sense that even though the exact location of a single molecule is not certain, the molecule could be spontaneously attracted to the so-called hot spots where the EM field is the strongest due to plasmon-induced optical forces.

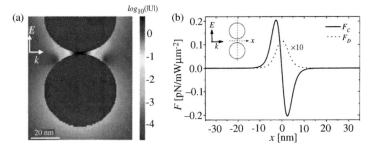

Figure 5.1 (a) Spatial optical potential distribution around a trimer of silver nanoparticles with 25 nm radius and 1 nm gap distance in water excited by the resonant wavelength (760 nm) at a power density of 10 mW/μm². (b) Optical force distribution at the gap of a dimer of Ag nanoparticles (radius 45 nm and gap distance 1 nm) along the incident wave vector (see inset) under 550 nm excitation. Reprinted with permission from Ref. [22]. Copyright (2002) by the American Physical Society.

For near-field-coupled metal nanoparticles, the optical force is enhanced because of the enhanced EM field gradient. This naturally leads to another interesting phenomenon: the spontaneous dimerization/aggregation of metal nanoparticles under external illumination [26, 32–34, 39–40]. Figure 5.2 shows the polar plot of the total optical force on the surface of the lower particle of the same dimer as in Fig. 5.1b. It is obvious that the force field is highly concentrated to the gap center, meaning that the particle itself experiences a net attractive force that would bring the two particles even closer. For comparison, the magnitude of the net force after surface integration is 62 pN under 1 mW/μm² toward the other particle in the dimer, while the radiation pressure force along the **k**-vector is only 0.3 pN.

From above, an attractive force is expected between spherical nanoparticles of the same size excited by a parallel polarized laser. However, a repulsive force is also theoretically predicted under certain excitation conditions for a dimer of two identical particles or heterodimers. Miljkovic et al. [33] calculated the optical force of a dimer of 5 nm Ag nanoparticles under both parallel (Fig. 5.3a) and perpendicular (Fig. 5.3b) polarizations. Although the two methods

(Maxwell's stress tensor and Mie theory [MST-Mie] shown by the thick solid lines and Lorentz force–coupled dipole approximation [LF-CDA] shown by the thin solid lines) produce slightly different results, the trends are the same. Under parallel polarization, an attractive force (negative values) is induced between the particles, in consistence with previously reported results. Interestingly, a repulsive force appears if the laser polarization is perpendicular to the dimer, although it is about 1 order of magnitude weaker than the attractive force in the former case. The repulsive force is easy to understand from a simple picture of dipole–dipole interaction. Neglecting retardation, two identical dipoles are in phase with each other under parallel polarization, leading to an attractive force. With perpendicular polarization, the two dipoles are out of phase, meaning that the sides with the same type of charges face each other, resulting in a repulsive force. A repulsive force can also be produced under parallel polarization for heterodimers, that is, a dimer of nonidentical spheres. In this case, the attractive and repulsive component of the optical force can even cancel out at certain separations and excitation wavelength, indicating that an equilibrium position exists between the particles [33]. One should note in Fig. 5.3 that Mie theory considers a large number of multipoles in the particles and is more accurate than the CDA method, which is dipole approximation based.

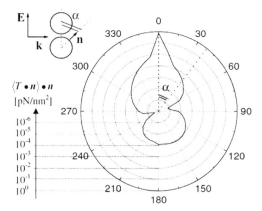

Figure 5.2 Polar plot of the optical force on the surface of the lower particle in a dimer under irradiation of 1 mW/µm^2 at 514.5 nm. Particle radius: 45 nm; gap distance: 1 nm. Reprinted with permission from Ref. [22]. Copyright (2002) by the American Physical Society.

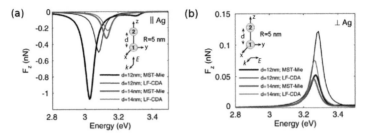

Figure 5.3 Spectra of the optical force between two identical silver spheres with radius $R = 5$ nm and interparticle distances $d = 12$ nm (black curves) and 14 nm (red curves) for the incident field of 1 W/μm² under parallel (a) and perpendicular (b) polarization configurations, respectively. The forces are calculated using two different methods: Maxwell's stress tensor and Mie theory (MST-Mie), and Lorentz force–coupled dipole approximation (LF-CDA). Reprinted with permission from Ref. [33]. Copyright (2010) American Chemical Society.

The above theoretical calculations assume that the structure of interest is under irradiation of a plane wave. From a practical point of view, a Gaussian beam is widely used in an optical tweezer system and thus would be more interesting to study theoretically. Li et al. [26] calculated the optical potential experienced by a metal nanoparticle in an optical trap of a focused laser beam in the presence of an immobilized one. The Gaussian beam was expanded into a sum of vector spherical harmonics (VSHs) with different orders by using the Davis formalism. Figure 5.4 shows the optical potential of a free silver nanoparticle of radius R as a function of the x position to the focus of a Gaussian beam, as illustrated in the inset, where F and I represent the free and immobilized particles, respectively, and D is the distance between the laser focus (the origin of the coordinate) and particle I. The incident laser wavelength is 514 nm for the green curves and 830 nm for the red curves. The results for particle radii 20 nm, 50 nm, and 100 nm are shown in Figs. 5.4a, 5.4b, and 5.4c, respectively. Taking the case of particle I being placed at the center of the beam ($D = 0$ nm, represented by the thin curves) as an example, it can be seen that for all the particle sizes, the optical potential of particle F drops dramatically when F approaches I, indicating that two particles can be easily dimerized in an optical trap of a focused laser beam. The left insets in the figure show the van der Waals potential between the two nanoparticles as a function of separation, which is comparable to the optical potential only for large particles at extremely short distances (e.g., 1–2 nm). The van der Waals force is an important factor that one has to consider while studying the interparticle forces in an optical trap, as will be shown in the following section.

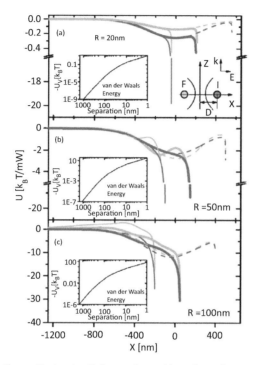

Figure 5.4 The optical potential experienced by a free silver particle *F* close to an identical immobilized particle *I* as a function of the *x* position. The right inset in (a) shows the optical trapping configuration for the two particles. The radii of spheres are R = 20, 50, and 100 nm for (a–c), respectively. Different separations, D = 0, 250, and 600 nm, between the beam focus and the position of the immobilized particle, are represented by thin, thick, and dashed lines (green for 514 nm excitation and red for 830 nm). The surrounding medium is water, and T = 300 K. The insets to the left in (a–c) are the corresponding van der Waals energies of two silver spheres as a function of the surface separation. Reprinted with permission from Ref. [26]. Copyright (2008) by the American Physical Society.

In general, the scattering force and absorption force point at the direction of the incident wave vector, as shown above, and have been considered detrimental for an optical trap, particularly in 3D trapping. Indeed, the absorption annihilates the incident photons and the resultant momentum is, of course, along the incident direction. However, the scattering component is more complicated. The magnitude and direction of the scattering force depend on the EM modes that are excited in a particle, so the scattering force could be used for trapping instead. Tractor beams that drag the particle moving against the incident wave vector have been theoretically demonstrated using shaped laser beams such as the Bessel beam [41, 42]. Even for a Gaussian beam, it is possible to produce an attractive scattering force on a metal nanoparticle [27]. For a gold nanoparticle (GNP) placed off-axis in the focal plane (see the inset in Fig. 5.5), if the wavelength of the incident laser is close to the quadrupole resonance, the interference between the dipole mode and the quadrupole mode results in stronger outward scattering than the inward scattering along the laser polarization, as shown in Fig. 5.5a. As a result, an attractive scattering force is exerted on the particle (Fig. 5.5b). Such an effect can in principle be used for size-selective 2D optical trapping in the focal plane. Figure 5.5c shows the calculated results for trapping wavelengths of 410 nm and 830 nm. It is obvious that the 410 nm laser produces an optical potential only for particles of radii around 60 nm and 100 nm and is not able to trap particles of other radii. For comparison, the 830 nm laser can trap particles of all radii, as shown in Fig. 5.5c.

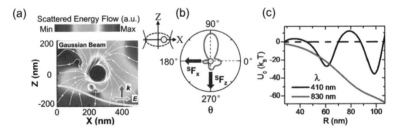

Figure 5.5 Attractive scattering force of a gold nanoparticle (radius 50 nm) excited by a Gaussian beam in water. (a) The scattering profile. The yellow curve represents the Gaussian profile of the incident intensity. The beam waist w_0 = 0.5 μm. The particle is positioned at $x = w_0/2$. The wave vector and polarization are indicated by red and green arrows. The white line and color scale are the direction and logarithmic modulus of the energy flow, respectively. (b) The corresponding scattering diagrams in x-z plane. The black arrows indicate the resulting scattering forces along x and z. (c) Potential well U_0 as a function of the particle radius under excitation of λ = 410 and 830 nm at 10 mW, indicating the size-selective particle sorting. The beam waist for both wavelengths is set to 0.5 μm. The environment is water, and T = 300 K. Reprinted with permission from Ref. [27]. Copyright (2014) American Chemical Society.

5.3 Experimental Demonstrations of Plasmon-Assisted Optical Forces

5.3.1 Optical Forces on Metal Nanoparticles Trapped by a Focused Laser Beam

Optical tweezers have been widely used to trap dielectric and metal nanoparticles. The distinct difference between dielectric and metal nanoparticles lies in the polarizability. Due to the larger polarizability, metal nanoparticles experience much higher optical forces, including the gradient, scattering, and absorption forces. On the other hand, the stable trapping of metal nanoparticles provides a platform for the study of physical properties of single metal nanoparticles and interactions between multinanoparticles. It also opens up the possibility of surface-enhanced spectroscopic sensing due to the near-field coupling effect between the trapped metal nanoparticles.

5.3.1.1 Elongated nanoparticles

Metal nanorods exhibit a polarizability difference along the long and short axes simply due to the different spatial confinement of the oscillation of free electrons. In an optical trap of a focused near-infrared (NIR) beam, the long axis of a metal nanorod turns out to self-align parallel to the laser polarization to minimize the potential energy of the induced dipole, given by $U = -\langle \mathbf{p} \cdot \mathbf{E} \rangle = -1/2 \sum \mathrm{Re}\{\alpha_{ii}\} E_i^2, i = x, y, z$, where $\mathbf{p}$ is the induced dipole moment, $\mathbf{E}$ is the incident electric field, and α_{ii} is the polarizability [43]. Furthermore, if the polarization of the trapping laser is rotated by an angle of β, the nanorod experiences a torque that equals $|\langle \mathbf{\tau} \rangle| = |\langle \mathbf{p} \times \mathbf{E} \rangle| = 1/4 E_0^2 (\alpha_{//} - \alpha_{\perp}) \sin 2\beta$ [44], where $\alpha_{//}$ and $\alpha_{\perp}$ are the polarizability components parallel and perpendicular to the long axis of the nanorod, respectively. This torque drives the nanorod to realign parallel to the laser polarization and rotate if the linear polarization is further rotated [43].

An example of a trapped Ag nanorod is shown in Fig. 5.6a. One sees a peak at ~500 nm in the dark-field (DF) scattering spectrum when the white-light polarization is parallel to the laser polarization and a peak at ~460 nm when the polarization is perpendicular. This indicates that the nanorod indeed aligns with its long axis parallel to the laser polarization. Figure 5.6b shows the polarizability plot for a silver nanorod of 60 nm diameter and 80 nm length in water as a function of the wavelength calculated using the modified long-wavelength approximation (MLWA). It is clearly seen that the polarizability of the nanorod is much higher at 830 nm, when the polarization of the laser is parallel rather than perpendicular to the long axis of the nanorod. Hence, a stronger dipole, and consequentially a deeper potential well, is induced if the long axis is parallel to the electric field of an 830 nm laser. Figure 5.6c demonstrates the rotation of the nanorod by rotating the linear polarization of the trapping laser. The polarization of white-light illumination is fixed while the polarization of the trapping laser is rotated up to 90°. It is seen from both the DF scattering images and spectra that the scattering intensity decreases when the laser polarization is rotated from parallel to perpendicular to the white-light polarization, implying that the nanorod is indeed rotated. It is worth noting that a dimer of spherical metal nanoparticles in principle behaves as a nanorod, with its dimer axis as the "long axis," and thus also aligns parallel to the laser polarization and can be rotated by rotating the laser polarization [40].

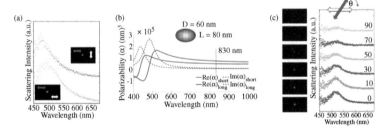

Figure 5.6 Alignment and rotation of a trapped Ag nanorod. Red and white arrows in (a) and (c) represent the polarization of the NIR trapping laser and white light, respectively. (a) Dark-field (DF) scattering spectra and images of a trapped nanorod. (b) Calculated polarizability of a nanorod (D = 60 nm and L = 80 nm). (c) DF images and spectra of a trapped nanorod at different angles between the laser polarization and the fixed white-light polarization. Reprinted with permission from Ref. [43]. Copyright (2010) American Chemical Society.

It has been previously reported that semiconductor nanowires of several microns in length can be trapped and manipulated to build nanostructures [45]. Shelton et al. [44] reported the optical trapping of glass nanowires in water and investigated the alignment and rotation behavior. They showed that a glass nanowire tends to align along the propagation of the trapping laser (514 nm) or parallel to the electric field if the self-alignment along the propagation is hindered. Silver nanowires in an optical trap also align themselves but, interestingly, perpendicular to the laser polarization [43, 46, 47]. In Fig. 5.7a, a silver nanowire of length ~5 μm is trapped at one of its tips (highlighted by the circle). The arrows on the right and left represent the polarization of the trapping laser and illuminating white light, respectively. Experimental observations show that, on the one hand, the preferential trapping spot is either of the ends of the nanowire, and on the other hand, more interestingly, the nanowire aligns perpendicular to the laser polarization. DDA simulations of the near fields generated by the wires under irradiation of an 830 nm plane wave were performed to understand the observations. The simulation results are shown in the right panel of Fig. 5.7b. For comparison, the near fields of a silver nanorod (D = 40 nm and L = 80 nm) are also given in the left panel. It is seen that, for a nanorod, the near fields are much stronger at parallel configuration than that at perpendicular configuration. However, the nanowire case is completely different. Much stronger near fields are induced when the laser polarization is perpendicular to the nanowire rather than parallel. This means that the nanowire tends to align perpendicularly to the laser polarization, which was observed experimentally. Another important feature is that the strongest field is always at the tips of the nanowire. This in principle explains the end-trapping of nanowires. Again, the rotation of the laser polarization produces an optical torque that drives the nanowire to rotate, as shown in Fig. 5.7a-I to 5.7a-III. The fixed nanowire to the left can be regarded as a reference.

By continuously rotating the linear polarization of the trapping laser, the trapped particles spin accordingly [43, 48, 49]. This is of particular interest from an application point of view. In a solution, due to the viscous hindrance of the surrounding medium, the axis of an elongated nanoparticle (a nanowire, a nanorod, or a dimer of spherical nanoparticles) is expected to lag behind the laser polarization. Such a phase delay strongly depends on the surface chemical properties of the particles, the viscosity of the surrounding medium, and the strength of the surface-plasmon-induced optical torque. Although the phase delay is difficult to observe on nanorods and dimers of spherical nanoparticles, perhaps due to the weaker drag force with respect to the surface-plasmon-induced torque, the lag is obvious in case of nanowires. For a nanowire of ~5 μm length and ~100 nm diameter, a phase delay of ~30° has been found at a rotation frequency of 0.8 Hz [43].

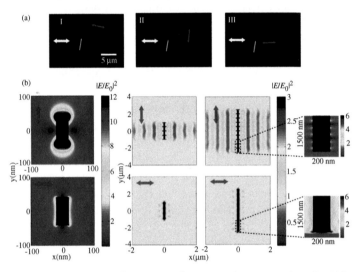

Figure 5.7 Alignment and rotation of a trapped Ag nanowire of ~100 nm in diameter and ~5 μm in length. (a) DF images of the trapped nanowire. Arrows on the right and left represent the laser polarization and white-light polarization, respectively. (b) Simulated near-field distribution of a nanorod (D = 40 nm and L = 80 nm) and two nanowires (D = 100 nm, L_1 = 2 μm, and L_2 = 5 μm) under irradiation of a plane wave of 830 nm. Reprinted with permission from Ref. [43]. Copyright (2010) American Chemical Society.

By applying a circular or elliptical polarization, the nanowires can even spin spontaneously. This is due to the direct transfer of the angular momentum from the incident photons to the nanowire. The spinning frequency is then proportional to the laser power at a specific polarization status. The direction of rotation can be easily changed by changing from right- to left-circularly-polarized light. One should point out that any anisotropic nanoparticles, such as nanorods and dimers of spherical nanoparticles, are also expected to spin spontaneously under irradiation of a circularly or elliptically polarized laser.

It is worth noting here that asymmetrical metal nanoparticles can also be rotated and spontaneously spin under illumination of a linearly polarized plane wave or a focused laser beam [31, 49]. Liu et al. [31] fabricated a gold nanomotor (100 nm radius, 380 nm line width, and 30 nm thickness) embedded by a silica microdisk that is 4000 times large in volume, which can spin when shined by a linearly polarized NIR laser. The direction of spinning can be changed simply by changing the laser wavelength. This is supported by their calculation of optical torque over 600–2400 nm wavelength range, which shows positive torques below ~1200 nm but negative values above.

5.3.1.2 Interaction between two metal nanoparticles in an optical trap

The understanding of the interactions between individual nanoparticles is crucial in the control of colloidal stability and in the rational guide of applications such as building blocks of self-assembled nanoparticles [50–52]. In an optical trap, two metal nanoparticles experience several forces that determine the response of the dimer. We will show that the surface-plasmon-induced optical force plays an important role in probing the physical properties of metal nanoparticles in a colloid at the single-particle level.

In a metal colloid, the particles are typically negatively charged due to the adsorption of anions, thus producing an electrostatic repulsion between adjacent particles [53]. This repulsive force depends on the number of surface charges and can be adjusted by changing the ionic strength of the solution. On the other hand, the van der Waals attractive force is determined by the material-specific Hamaker constant and particles' sizes and separation. The two forces, the Coulomb repulsion and van der Waals attraction, are the main components of the well-known Derjaguin–Landau–Verwey–Overbeek (DLVO) forces, which are used to interpret the colloidal stability [53–55]. The total potential is then the sum of DLVO and interparticle optical potentials.

In Fig. 5.8, two GNPs of diameter 80 nm were trapped and dimerized in an optical trap of a NIR laser (830 nm) [40]. Figures 5.8a and 5.8b show the DF scattering images and spectra, respectively. The final concentration of salt (NaCl) in the colloid is adjusted to be 20 mM. The white and red double arrows represent the polarization of white light for DF illumination and the trapping laser, respectively. Figures 5.8a-I and 5.8a-II are the DF images of a single GNP. They show that the scattering color does not change with the change of white-light polarization, indicating the spherical shape of the particle. However, the DF scattering of a dimer of particles shows clear polarization dependence, as seen in Fig. 5.8a-III and 5.8a-IV. If the laser polarization is parallel to the white-light polarization, the surface plasmon resonance peaks at ~620 nm due to the near-field coupling and the dimer is seen as reddish in DF imaging (Fig. 5.8a-III). If the polarizations are perpendicular (Fig. 5.8a-IV), the resonance peak is at ~560 nm, which is close to the resonance of a single particle. An interesting conclusion can be drawn from Figs. 5.8a and 5.8b that the dimer of spherical nanoparticles also aligns parallel to the laser polarization, similar to the nanorod case that is shown previously. Figure 5.8c shows the calculated potential curves of the dimer with (lower curve) and without (upper curve) optical potential. The dominant features in the curves are a deep primary minimum at near-zero separation and a potential barrier that prevents irreversible aggregation. Another important feature is that there is a shallow secondary minimum at ~15 nm separation without optical potential. For some colloids and ionic strengths, this secondary minimum may result in spontaneous aggregation or flocculation [56], but here the depth is ~1 k_BT and not deep enough to induce stable dimerization against the thermal Brownian motion. Consequently, the DLVO potential curve indicates that such a colloid should be stable against aggregation up to a salt concentration of at least 20 mM. However, as illustrated by the lower curve in Fig. 5.8c, the secondary minimum deepens dramatically if the laser-induced attractive optical potential is added. For a laser power of only 2 mW/μm² at λ = 830 nm, a potential well depth of ~5 k_BT at ~9 nm separation was obtained. This leads to stable dimerization and a significant red shift in DF scattering due to strong plasmon hybridization (see Fig. 5.8b).

For a colloid, adding salt leads to an increased screening of the surface charges through a build-up of the density of counterions on the particle surfaces and within their immediate surroundings, the so-called electrical double layer [53, 57]. As a result, the electrostatic repulsion should decrease with increasing salt concentration and the position of the secondary potential minimum should move toward smaller separations, leading to an increased near-field coupling and spectral red shift for a dimer of GNPs. Such a concentration-dependent red shift is clearly shown in Fig. 5.9a. The measured peak positions (circles) of the longitudinal mode of the dimers are plotted for different NaCl concentrations of 0, 5, 10, 15, and 20 mM. On the other hand, the perpendicular plasmon mode remains close to the single-particle resonance position.

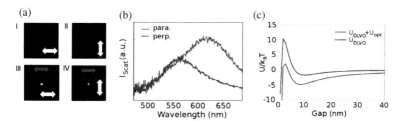

Figure 5.8 (a) DF images of two single gold nanoparticles of 80 nm before and after dimerization with white-light polarization (white arrows) parallel and perpendicular to the laser polarization (top arrows in images III and IV). (b) Corresponding DF scattering spectra. (c) Calculated total potentials of the dimer in a colloid with and without optical potential. NaCl concentration: 20 mM. Reprinted with permission from Ref. [40]. Copyright (2011) American Chemical Society.

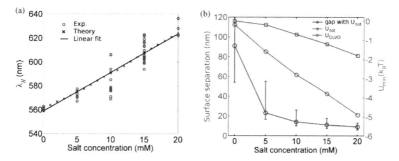

Figure 5.9 (a) Experimental and simulated peak positions of the longitudinal mode ($\lambda_{//}$) as a function of the salt concentration. The solid line is a linear fit to the simulated results (crosses). (b) Simulated surface separations and depth of U_{min} at different salt concentrations. The error bars represent the separation ranges corresponding to a variation of 1 $k_B T$ centered at U_{min} in the potential well. Reprinted with permission from Ref. [40]. Copyright (2011) American Chemical Society.

From a theoretical point of view, the total potential can be approximated as $U_{tot} = U_c + U_{vdW} + U_{opt}$, where the first two terms, the Coulomb repulsion potential and van der Waals attraction potential, constitute the DLVO potential and the third is the optical potential. The optical potential U_{opt} is the sum of two terms U_{inter} and U_{trap}, that is, the potentials due to interparticle coupling and particle confinement within the laser focus, respectively. The latter term turns out to be much smaller than the interparticle potential and can be safely neglected. The former item, U_{inter}, can be calculated using the Maxwell's stress tensor formalism combined with Mie theory as described above in the theoretical calculations section [33].

The theoretical scattering spectra can be obtained from the numerical integration of the product of normalized probability distribution and the calculated spectra at a certain separation. The peak positions can then be found and are shown by the crosses in Fig. 5.9a together with a linear fit (black line), which finds an excellent agreement with the experimental results. The minimum most-probable separation of the two particles, as well as the depth of the secondary minimum potential well, can also be obtained and are shown in Fig. 5.9b. Although there is a considerable variation in the red shift between different particle pairs measured under identical external conditions, probably due to the difference in surface charge densities of particles, the methodology described herein provides a valuable tool for characterizing nanoparticle heterogeneities that are not accessible through ensemble averaged measurements.

5.3.1.3 Applications in SERS sensing

As mentioned in the previous sections, optical forces induce spontaneously aggregation of metal nanoparticles, which in turn leads to strong near-field coupling and produces high field enhancement in the hot spots. Such aggregates are excellent candidates for SERS applications.

A dimer of two metal nanoparticles created optically in a focused laser beam has shown high enough field enhancement for efficient SERS detection [39]. The example in Fig. 5.10a shows the dimerization of two silver nanoparticles using optical tweezers to create hot spots for SERS. In these experiments, a NIR laser of 830 nm is used for optical trapping and a separate laser at 514.5 nm is used for Raman excitation. Thiophenol molecules are chemically bound to the surface of the silver nanoparticles as Raman probe. A single silver nanoparticle immobilized on a glass slide or trapped by the focused laser can be identified by DF imaging as a bluish spot, from which no SERS signal is detected (insets in the top and middle panels). However, if a dimer is created by moving a trapped particle to an immobilized one, a significant SERS signal arises due to the strong near-field coupling effect (see the inset in the bottom panel).

Figure 5.10b shows the calculated optical potential experienced by a silver nanoparticle that is free to move in a Gaussian laser focus at 830 nm wavelength and 50 mW laser power. The white circle represents an immobilized particle that is 250 nm away from the laser focus along the x axis. The result is similar to that shown in Fig. 5.4, where a deep potential minimum is induced when the trapped particle approaches the immobilized one, giving rise to spontaneous optical dimerization and a SERS hot spot in the optical trap, which is responsible for the appearance of a SERS signal after dimerization.

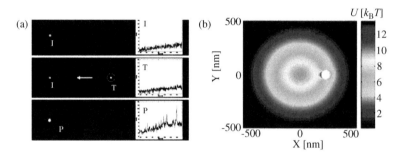

Figure 5.10 (a) The dimerization of silver nanoparticles, DF images, and corresponding Raman spectra. (b) Simulated optical potential under 830 nm laser irradiation experienced by a trapped Ag nanoparticle with an immobilized one that is 250 nm away from the laser focus. Reprinted with permission from Ref. [39]. Copyright (2006) American Chemical Society.

For larger aggregates, stronger enhancement is expected [58]. The work by Tanaka et al. [59] showed that multiple silver nanoparticles simultaneously trapped by a NIR laser beam also tend to align parallel to the linear polarization and that, more importantly, the aggregates exhibited pronounced SERS signal at a very low concentration of probe molecules, 10^{-14} M pseudoisocyanine (PIC), excited by a separate laser beam, whereas no SERS appeared if the trapping laser was not focused in the solution.

Optical-force-induced aggregation in laser tweezers has prominent potential for lab-on-a-chip-based chemo-/biosensing applications [60]. In a microfluidic channel, the metal nanoparticles incubated by the analyte molecules can be trapped in the flow and form aggregates, which are ideal substrates for SERS sensing. Figure 5.11a shows DF images of the growing aggregate created by optical tweezers using NIR laser trapping at 830 nm. The corresponding SERS spectra, excited by a 514.5 nm laser line, are shown in Fig. 5.11b. With a growth in size, the aggregate shows increased SERS enhancement. Using microfluidics, the analyte flow is interchangeable, enabling consecutive detection of a series of species simply by changing the analyte solutions and keeping the Ag colloid flow. With a Y-shaped microfluidic channel, where metal colloid and analyte solution can flow in parallel, a SERS signal can also be detected at the interface where the nanoparticles and the analyte molecules are mixed through diffusion [60].

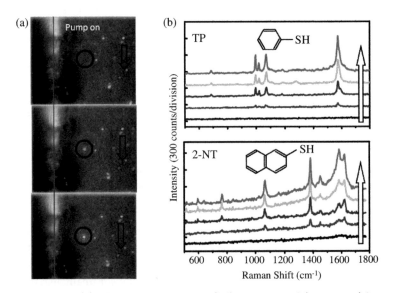

Figure 5.11 (a) DF scattering images of silver nanoparticles trapped in a microfluidic channel. The trapping wavelength was 830 nm. (b) Corresponding SERS spectra from the optically induced nanoparticle aggregates. A separate laser line at 514.5 nm was used to excite the Raman probes. Adapted from Ref. [60] with permission of The Royal Society of Chemistry.

5.3.2 Optical Forces in Lithographically Fabricated Plasmonic Nanostructures

By using metal nanostructures fabricated on a substrate, plasmon-assisted optical forces can be used to manipulate objects in a more controllable way [30, 61–65]. An advantage of such experiments is that the laser power density can be much lower than that used in optical tweezers and thus would cause a much less heating effect resulting from absorption and less damage in biological applications.

5.3.2.1 Gold nanopads and nanoholes

For gold nanopads fabricated on a glass substrate, if excited by total internal reflection illumination, the optical potential experienced by a dielectric nanoparticle exhibits a potential well centered at the forward edges of the nanopads [61, 62]. Figure 5.12a shows the map of optical potential of a 200 nm polystyrene (PS) bead located 20 nm above a 450 nm × 450 nm gold pad of 40 nm thickness under 785 nm laser irradiation of power density 5×10^7 W/m². The asymmetrical distribution of the potential mainly arises from the strong forward scattering due to the illumination asymmetry. Nevertheless, the depth of the potential well is stable enough to trap the PS bead, as demonstrated in Fig. 5.12b, where arrays of patterned gold nanopads are used.

Besides metal nanodisks, nanoholes in a metal film can also trap dielectric beads under normal illumination at resonance. It has been shown that PS beads of 50 nm and 100 nm can be trapped in cylindrical apertures of a diameter of a few hundred nanometers in a 100 nm thick gold film using a 1064 nm laser [65]. The laser power is less than 2 mW, significantly lower than that used in conventional optical trapping experiments with laser tweezers. Figure 5.12c shows the scheme of the experiments. As shown in Fig. 5.12d, the transmitted intensity of the laser fluctuates during the trapping process. If a PS nanoparticle is trapped at the edge of an aperture, the transmission increases. Although the particle tends to escape due to the thermal motion, the displacement of the bead would result in a restoring force that pushes it back, which the authors called self-induced back-action (SIBA).

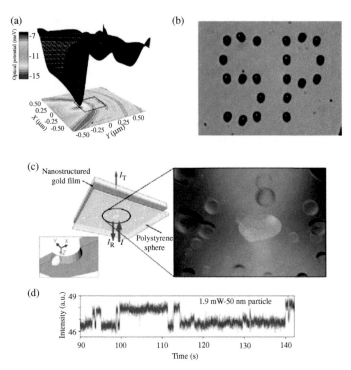

Figure 5.12 (a) Theoretical optical potential experienced by a 200 nm PS bead placed 20 nm above a gold nanopad. (b) PS beads trapped on a patterned array of gold nanopads. (c) Scheme of optical trapping using nanoholes in a gold film. (d) Intensity fluctuation of the laser focused on a nanohole due to the optical trapping of a PS bead. Reprinted by permission from Macmillan Publishers Ltd: [*Nature Physics*] (Refs. [62, 65]), copyright (2007, 2009).

5.3.2.2 Dimers of nanodisks

The coupling between metal nanostructures produces extra field enhancement in the gaps [66, 67], which consequently enables more precise optical trapping and positioning of nanoscale objects. Compared to a conventional optical tweezer setup, the trapping efficiency can be greatly improved and the Brownian motion of the trapped object can be suppressed by almost 1 order of magnitude [30, 63]. Figure 5.13a shows the recorded positions of a trapped single PS bead with and without the presence of a gold nanodimer (time step: 5 ms). In conventional 3D optical tweezers without gold nanodisks (top-left scheme), the movement of the PS bead is confined within the trap created by the laser itself, which is a circle of ~1 µm diameter (green solid circles). However, if the trapped bead is close to a dimer of gold nanodisks (top-right scheme), it experiences a strong local electric field and is attracted to the gap. The attractive force results in strong confinement of the bead and thus a greatly suppressed Brownian motion, as shown by the red circles in Fig. 5.13a. The full-width at half-maximum (FWHM) of the position distribution can be reduced by approximately 10 times from ~400 nm to ~40 nm [63].

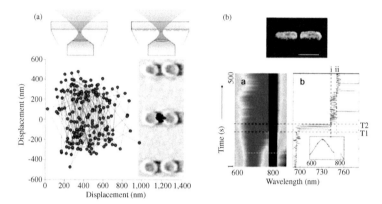

Figure 5.13 Near-field trapping of PS nanobeads (a) and gold nanoparticles (b) using a dimer of gold nanodisks (a) and nanocuboids (b). Scale bar in the upper panel of (b): 100 nm. (a) Reprinted by permission from Macmillan Publishers Ltd: [*Nature Photonics*] (Ref. [63]), copyright (2008). (b) Reprinted with permission from Ref. [30]. Copyright (2010) American Chemical Society.

When a metal nanoparticle is close to the gap of a dimer of gold nanocuboid, strong near-field coupling occurs between the metal nanoparticle and the nanoantenna, which produces extra near fields and enhances the trapping effect. Zhang et al. [30] reported the trapping of 10 nm GNPs at the gap of a gold dipole nanoantenna with a 5 to 30 nm gap size (a typical scanning electron microscopy [SEM] image shown in the upper panel in Fig. 5.13b) using a slightly focused 808 nm laser. Individual trapping events are monitored in real time by the Rayleigh scattering of a single antenna. As shown in the lower panel of Fig. 5.13b, when a GNP was trapped, a sudden red shift appeared due to the coupling between the particle and the antenna. The trapping behavior is also shown to be reversible, that is, by turning the trapping laser off, the trapped particle escapes from the gap.

5.3.3 Optical Forces in Propagating Surface Plasmon Systems: Gold Thin Films and Nanostripes

Since optical force in plasmonic systems originates from the near fields, it exists not only in metal nanostructures that confine surface plasmons but also in metal thin films that support propagating surface plasmons [24, 25, 38, 68]. The difference is that in surface plasmon polariton (SPP) systems, the scattering force always pushes objects to move along the propagation. However, under counterbeam excitation or normal illumination, the propulsion can be canceled out and stable trapping is obtainable.

Figure 5.14 shows the measured force vectors (left) and magnitudes (Fig. 5.14b, triangles: z component; squares: x component; and circles: module) of a 2 μm dielectric bead as a function of the distance from a gold film (40 nm thickness) under 633 nm excitation at the plasmon resonance angle (~71° in this example) [24]. The z component is from the gradient of the evanescent field, and the x component is along the **k**-vector and is a result of plasmon propagation. It is clear that both components increase with decreasing distance, but the x component increases more rapidly than the z component, meaning that the dielectric bead will be pushed toward the plasmon propagation direction when it approaches the gold film. The right panel in Fig. 5.14 shows the optical force of beads of different sizes at constant surface separation (500 nm) from the film. Smaller particles experience larger forces, and the direction of the force vector is different for different particles. This effect could in principle be applied for optical sorting of nanoparticles.

In the case of metallic nanoparticles, the propulsion is further enhanced due to the near-field coupling between the particle and the film [68]. With counterpropagating SPPs excited by two incident beams at a specific incident angle, the propulsion from the two beams can cancel out to achieve stable trapping. Wang et al. [25] reported stable trapping of a dielectric bead (1 μm PS particle) on a gold stripe of 1 or 2 μm width. The experimental setup is shown in Fig. 5.15a. Figure 5.15b shows the optical images of a PS bead (the bright spot in the center of the top-left image) propelled by a single laser beam. It is clearly seen that the particle is pushed to move toward the left side along the gold stripe. By using two counterbeams, the particle can be held stably at an equilibrium position in the middle determined by the intensities of the two incident beams. In Fig. 5.15c, the position of a PS bead was recorded within 1 min at 30 measurements per second on a stripe of 1 μm width with equal laser intensity from both sides. It is seen that the particle is well confined along the *x* direction due to the electric field confinement, as shown in Fig. 5.15d.

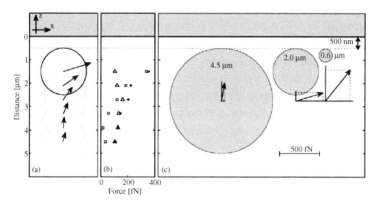

Figure 5.14 Optical force experienced by a dielectric bead close to a 40 nm thick gold thin film. (a) Force vectors as a function of the distance of a 2 μm sphere from the surface of the gold film. (b) *x* (squares) and *z* (triangles) components and module (circles) of the force. (c) Force vectors of beads of different sizes at a 500 nm separation from the gold film. Reprinted with permission from Ref. [24]. Copyright (2006) by the American Physical Society.

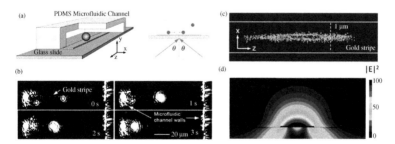

Figure 5.15 (a) Schematic of the experimental design. (b) Optical images of a PS bead pushed to move along the gold nanostripe. (c) Spatial positions of a PS bead on the gold nanostripe of 1 μm width illuminated from both left and right sides. (d) Simulated electric field component $|E|^2$ of first-order quasi-TM mode on the gold stripe (1 μm wide and 40 nm thick). Reprinted with permission from Ref. [25]. Copyright (2010) American Chemical Society.

5.4 Summary and Perspective

The plasmonic feature of metal nanostructures opens an avenue for nanomanipulation of objects by means of light. When subjected to a laser tweezer, metal nanoparticles exhibit dramatically different behaviors compared to dielectric particles due to the excitation of surface plasmons. The unique optical force results in particular alignment and rotation of metal nanoparticles, for example, nanorods align parallel to the polarization of a NIR trapping laser, while nanowires align perpendicularly. The spontaneous dimerization and aggregation enable the investigation of interparticle interactions at a single-particle level and show important applications in lab-on-a-chip-based surface-enhanced spectroscopic sensors. Combined with prefabricated metal nanostructures on a substrate, surface-plasmon-assisted optical forces also enable highly precise manipulation that is beyond the diffraction limit due to the localization of the strong near fields. It lowers the requirements for optical trapping of nanoscale objects; in particular, the laser intensity can be significantly reduced, facilitating noninvasive manipulation of nanoscale objects, for example, biological molecules/cells. Although it has been studied extensively, the research of surface-plasmon-assisted optical forces still requires further investigation into electrodynamics calculation methods, for example, for more complex systems such as coupled nanoparticles/nanowires, and it would find more interesting applications such as the construction of networks of metal nanostructures for optical circuit devices.

References

1. D. G. Grier, *Nature*, **424**, 810 (2003).
2. M. Dienerowitz, M. Mazilu, and K. Dholakia, *J. Nanophotonics*, **2**, 021875 (2008).
3. M. L. Juan, M. Righini, and R. Quidant, *Nat. Photonics*, **5**, 349 (2011).
4. A. Jonas, and P. Zemanek, *Electrophoresis*, **29**, 4813 (2008).
5. K. Dholakia, P. Reece, and M. Gu, *Chem. Soc. Rev.*, **37**, 42 (2008).
6. C. J. Min, Z. Shen, J. F. Shen, Y. Q. Zhang, H. Fang, G. H. Yuan, L. P. Du, S. W. Zhu, T. Lei, and X. C. Yuan, *Nat. Commun.*, **4**, 2891 (2013).
7. A. Ashkin, *Phys. Rev. Lett.*, **24**, 156 (1970).

8. A. Ashkin, J. M. Dziedzic, J. E. Bjorkholm, and S. Chu, *Opt. Lett.*, **11**, 288 (1986).
9. Y. Arai, R. Yasuda, K. Akashi, Y. Harada, H. Miyata, K. Kinosita, Jr., and H. Itoh, *Nature*, **399**, 446 (1999).
10. J. Prikulis, F. Svedberg, M. Käll, J. Enger, K. Ramser, M. Goksor, and D. Hanstorp, *Nano Lett.*, **4**, 115 (2004).
11. P. M. Hansen, V. K. Bhatia, N. Harrit, and L. Oddershede, *Nano Lett.*, **5**, 1937 (2005).
12. J. R. Moffitt, Y. R. Chemla, S. B. Smith, and C. Bustamante, *Annu. Rev. Biochem.*, **77**, 205 (2008).
13. D. J. Stevenson, F. Gunn-Moore, and K. Dholakia, *J. Biomed. Opt.*, **15**, 041503 (2010).
14. K. Svoboda, and S. M. Block, *Opt. Lett.*, **19**, 930 (1994).
15. J. Do, M. Fedoruk, F. Jackel, and J. Feldmann, *Nano Lett.*, **13**, 4164 (2013).
16. M. Moskovits, *Rev. Mod. Phys.*, **57**, 783 (1985).
17. K. Kneipp, Y. Wang, H. Kneipp, L. T. Perelman, I. Itzkan, R. Dasari, and M. S. Feld, *Phys. Rev. Lett.*, **78**, 1667 (1997).
18. S. M. Nie, and S. R. Emery, *Science*, **275**, 1102 (1997).
19. K. A. Willets, and R. P. Van Duyne, *Annu. Rev. Phys. Chem.*, **58**, 267 (2007).
20. J. N. Anker, W. P. Hall, O. Lyandres, N. C. Shah, J. Zhao, and R. P. Van Duyne, *Nat. Mater.*, **7**, 442 (2008).
21. H. X. Xu, J. Aizpurua, M. Käll, and P. Apell, *Phys. Rev. E*, **62**, 4318 (2000).
22. H. X. Xu, and M. Käll, *Phys. Rev. Lett.*, **89**, 246802 (2002).
23. N. Calander, and M. Willander, *Phys. Rev. Lett.*, **89**, 143603 (2002).
24. G. Volpe, R. Quidant, G. Badenes, and D. Petrov, *Phys. Rev. Lett.*, **96**, 238101 (2006).
25. K. Wang, E. Schonbrun, P. Steinvurzel, and K. B. Crozier, *Nano Lett.*, **10**, 3506 (2010).
26. Z. P. Li, M. Käll, and H. X. Xu, *Phys. Rev. B*, **77**, 085412 (2008).
27. Z. P. Li, S. P. Zhang, L. M. Tong, P. J. Wang, B. Dong, and H. X. Xu, *ACS Nano*, **8**, 701 (2014).
28. Y. G. Song, B. M. Han, and S. Chang, *Opt. Commun.*, **198**, 7 (2001).
29. R. Quidant, and C. Girard, *Laser Photonics Rev.*, **2**, 47 (2008).
30. W. H. Zhang, L. N. Huang, C. Santschi, and O. J. F. Martin, *Nano Lett.*, **10**, 1006 (2010).

31. M. Liu, T. Zentgraf, Y. M. Liu, G. Bartal, and X. Zhang, *Nat. Nanotechnol.*, **5**, 570 (2010).
32. A. J. Hallock, P. L. Redmond, and L. E. Brus, *Proc. Natl. Acad. Sci. USA*, **102**, 1280 (2005).
33. V. D. Miljkovic, T. Pakizeh, B. Sepulveda, P. Johansson, and M. Käll, *J. Phys. Chem. C*, **114**, 7472 (2010).
34. R. A. Nome, M. J. Guffey, N. F. Scherer, and S. K. Gray, *J. Phys. Chem. A*, **113**, 4408 (2009).
35. E. Lamothe, G. Leveque, and O. J. F. Martin, *Opt. Express*, **15**, 9631 (2007).
36. F. J. G. de Abajo, T. Brixner, and W. Pfeiffer, *J. Phys. B: At. Mol. Opt. Phys.*, **40**, S249 (2007).
37. R. Quidant, D. Petrov, and G. Badenes, *Opt. Lett.*, **30**, 1009 (2005).
38. D. Woolf, M. Loncar, and F. Capasso, *Opt. Express*, **17**, 19996 (2009).
39. F. Svedberg, Z. P. Li, H. X. Xu, and M. Käll, *Nano Lett.*, **6**, 2639 (2006).
40. L. M. Tong, V. D. Miljkovic, P. Johansson, and M. Käll, *Nano Lett.*, **11**, 4505 (2011).
41. N. Wang, J. Chen, S. Y. Liu, and Z. F. Lin, *Phys. Rev. A*, **87**, 063812 (2013).
42. A. Novitsky, C. W. Qiu, and H. F. Wang, *Phys. Rev. Lett.*, **107**, 203601 (2011).
43. L. M. Tong, V. D. Miljkovic, and M. Käll, *Nano Lett.*, **10**, 268 (2010).
44. W. A. Shelton, K. D. Bonin, and T. G. Walker, *Phys. Rev. E*, **71**, 036204 (2005).
45. P. J. Pauzauskie, A. Radenovic, E. Trepagnier, H. Shroff, P. D. Yang, and J. Liphardt, *Nat. Mater.*, **5**, 97 (2006).
46. Z. Yan, M. Pelton, L. Vigderman, E. R. Zubarev, and N. F. Scherer, *ACS Nano*, **7**, 8794 (2013).
47. Z. Yan, J. E. Jureller, J. Sweet, M. J. Guffey, M. Pelton, and N. F. Scherer, *Nano Lett.*, **12**, 5155 (2012).
48. K. D. Bonin, B. Kourmanov, and T. G. Walker, *Opt. Express*, **10**, 984 (2002).
49. P. H. Jones, F. Palmisano, F. Bonaccorso, P. G. Gucciardi, G. Calogero, A. C. Ferrari, and O. M. Marago, *ACS Nano*, **3**, 3077 (2009).
50. Z. Y. Tang, N. A. Kotov, and M. Giersig, *Science*, **297**, 237 (2002).
51. A. Yethiraj, and A. van Blaaderen, *Nature*, **421**, 513 (2003).
52. T. P. Bigioni, X. M. Lin, T. T. Nguyen, E. I. Corwin, T. A. Witten, and H. M. Jaeger, *Nat. Mater.*, **5**, 265 (2006).

53. D. J. Shaw, *Introduction to Colloid and Surface Chemistry*, 4th ed. (Butterworth-Heinemann, 2000).
54. J. Lyklema, H. P. van Leeuwen, and M. Minor, *Adv. Colloid Interface Sci.*, **83**, 33 (1999).
55. J. N. Israelachvili, *Intermolecular & Surface Forces*, 2nd ed. (Academic Press Limited, 1992).
56. M. W. Hahn, D. Abadzic, and C. R. O'Melia, *Environ. Sci. Technol.*, **38**, 5915 (2004).
57. J. E. Sader, S. L. Carnie, and D. Y. C. Chan, *J. Colloid Interface Sci.*, **171**, 46 (1995).
58. F. Svedberg, and M. Käll, *Faraday Discuss.*, **132**, 35 (2006).
59. Y. Tanaka, H. Yoshikawa, T. Itoh, and M. Ishikawa, *J. Phys. Chem. C*, **113**, 11856 (2009).
60. L. M. Tong, M. Righini, M. U. Gonzalez, R. Quidant, and M. Käll, *Lab Chip*, **9**, 193 (2008).
61. M. Righini, C. Girard, and R. Quidant, *J. Opt. A: Pure Appl. Opt.*, **10**, 093001 (2008).
62. M. Righini, A. S. Zelenina, C. Girard, and R. Quidant, *Nat. Phys.*, **3**, 477 (2007).
63. A. N. Grigorenko, N. W. Roberts, M. R. Dickinson, and Y. Zhang, *Nat. Photonics*, **2**, 365 (2008).
64. Z. Y. Fang, F. Lin, S. Huang, W. T. Song, and X. Zhu, *Appl. Phys. Lett.*, **94**, 063306 (2009).
65. M. L. Juan, R. Gordon, Y. J. Pang, F. Eftekhari, and R. Quidant, *Nat. Phys.*, **5**, 915 (2009).
66. L. Gunnarsson, E. J. Bjerneld, H. X. Xu, S. Petronis, B. Kasemo, and M. Käll, *Appl. Phys. Lett.*, **78**, 802 (2001).
67. S. K. Ghosh, and T. Pal, *Chem. Rev.*, **107**, 4797 (2007).
68. K. Wang, E. Schonbrun, and K. B. Crozier, *Nano Lett.*, **9**, 2623 (2009).

Chapter 6

Plasmonic Nanowire Waveguides and Circuits

Hong Wei[a] and Hongxing Xu[b]

[a]*Institute of Physics, Chinese Academy of Sciences, Beijing 100190, China*
[b]*School of Physics and Technology, and Institute for Advanced Studies,
Wuhan University, Wuhan 430072, China*
weihong@iphy.ac.cn; hxxu@whu.edu.cn

6.1 Introduction

The development of nanoelectronic integrated circuits is approaching a bottleneck due to the limited operating speed and high power consumption. Compared to electrons, photons as information carriers have the advantage of fast speed, large bandwidth, and low power consumption, which can be used for developing new information technology. However, due to the diffraction limit of light, the scale of dielectric photonic devices is large, which limits the on-chip integration of photonic devices. Since surface plasmon polaritons (SPPs) can concentrate electromagnetic energy into deep subwavelength volumes, SPP-based devices and components can break the diffraction limit of light and scale down to the nanometer scale. Hence, they are promising for building nanophotonic

---

Nanophotonics: Manipulating Light with Plasmons
Edited by Hongxing Xu
Copyright © 2018 Pan Stanford Publishing Pte. Ltd.
ISBN 978-981-4774-14-7 (Hardcover), 978-1-315-19661-9 (eBook)
www.panstanford.com

integrated circuits and merging photonics and electronics at the nanoscale [1, 2].

One critical building block to construct a plasmonic circuit is the plasmonic waveguide, which can be regarded as a nanoscale counterpart of an optical fiber that guides light with subwavelength confinement of the optical field. Although long-range SPPs have been proposed a long time ago for electromagnetic energy transport [3], their lack of subwavelength field localization makes them unsuitable for the ultracompact integrated circuits. Starting from late 1990s, most attention has been devoted to the plasmon modes with highly localized fields in 1D structures. In 1997, Takahara et al. theoretically proposed the 1D optical waveguide with subwavelength field confinement [4]. In 1998, Quinten et al. theoretically investigated the electromagnetic energy transport in nanoparticle chains [5]. The strong energy dissipation makes this kind of waveguide only support plasmon propagation at a very short distance of the order of one micron or less [6, 7]. Metal strips are one of the most studied plasmonic waveguides [8–11]. The SPP propagation in metal strips was experimentally investigated by using a scanning near-field optical microscope (SNOM) [12–14]. Grooves in a metal film have been investigated theoretically and experimentally as waveguides supporting channel plasmon polaritons [15–17]. In addition, wedges, slots, and dielectric waveguides over metal films have also been proposed as plasmonic waveguides [18–23].

Besides the above-mentioned waveguides, another kind of waveguide is metal (silver and gold) nanowires (NWs), which are usually prepared by bottom-up chemical methods. Due to the crystalline structure of the chemically synthesized metal NWs, the damping of SPPs is low during the propagation. Moreover, these NWs can be manipulated conveniently and are good media to transmit electric signals. These characteristics make this kind of NWs a good candidate for both the study of fundamental nanophotonic sciences and the proof-of-concept demonstration of potential applications. In experiments, the NWs used as plasmonic waveguides are most silver NWs because they can support SPP propagation over longer distances in the visible and near-infrared spectral range. In this chapter, we will focus on metal NWs and simple networks for the construction of plasmonic integrated circuits. This chapter is organized as follows: First in Section 6.2, the experimental aspects about surface plasmon excitation and detection are introduced. Section 6.3 presents the basic properties of surface plasmons in the NW waveguides, including SPP modes, propagation, group velocity, loss, emission, spin-dependent properties, and NW–emitter interactions. Section 6.4 introduces the plasmonic devices and circuits made of metal NWs, including SPP routers and demultiplexers, SPP-interference-based logic gates, and hybrid plasmonic-photonic NW systems. Finally in Section 6.5, we briefly summarize this chapter.

6.2 Excitation and Detection of Propagating SPPs

6.2.1 SPP Excitation

For the SPPs at the flat interface between metal and dielectric half spaces, the wave vector of SPPs is expressed as

$$k_{spp} = k_0 \sqrt{\frac{\varepsilon_d \varepsilon_m}{\varepsilon_d + \varepsilon_m}}, \quad (6.1)$$

where k_0 is the free-space light wave vector equal to ω/c (ω and c are the light frequency and speed, respectively) and ε_d and ε_m are frequency-dependent permittivity of dielectric and metal materials, respectively. This expression defines the dispersion relation of SPPs. Since k_{spp} is larger than k_0, the SPPs cannot be excited directly by free-space light. Some excitation schemes are needed to guarantee the momentum matching of SPPs and photons. To excite the propagating SPPs in metal NWs using light, the momentum matching condition also needs to be fulfilled.

One method to match the wave vector of light and SPPs is using the Kretschmann–Raether configuration (also known as Kretschmann configuration), in which a glass prism is used [24]. The metal NW is deposited on top of the prism, and the excitation light is incident onto the prism with an angle at which the light wave vector component parallel to the NW matches the SPP wave vector, so the SPPs propagating along the NW can be excited (Fig. 6.1a). Usually focused laser light is incident through the prism onto one end of the NW waveguide [25].

The light scattering at the discontinuous points (e.g., NW ends and sharp kinks) on the NW is another way to excite SPPs. If focused light is incident on these points, SPPs will be launched (Fig. 6.1b) [26]. Nanoparticles adhered to the NWs can also provide the scattering to excite the plasmons on the wires [27]. The NW end and the adhered nanoparticle function as optical antennas to covert freely propagating light into SPPs. To improve the coupling efficiency, well-designed nanoantennas can be fabricated at the NW ends as efficient couplers [28, 29].

If an optical fiber is pulled to get a sharp taper with the size comparable to the NW waveguide, the tapered fiber tip approaching the NW waveguide can excite the SPPs on the NW through evanescent field coupling (Fig. 6.1c) [30–33]. The optical fiber excitation scheme also provides a possible way to combine the dielectric optics and plasmon-based nano-optics. The convenient manipulation of the optical fiber and the free excitation sites on the NW make it a very useful excitation method.

The decay of excited quantum emitters, such as atoms, molecules, quantum dots (QDs), and nitrogen vacancy (NV) centers in nanodiamonds, can excite SPPs on the NW. Figure 6.1d shows the sketch and experimental fluorescence image of exciting SPPs on a Ag NW by illuminating a QD. The bright spot A in the fluorescence image is the QD emission, while the spots B and C are the scattering of the QD-excited SPPs at the NW ends. The single-photon emission of the QD generates single quantized plasmons in the NW.

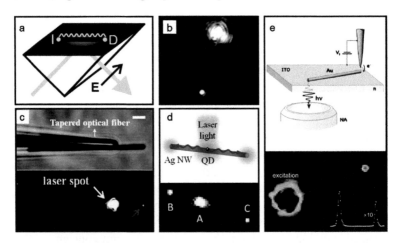

Figure 6.1 Excitation methods for SPPs propagating in metal NWs. (a) Reprinted with permission from Ref. [34]. Copyright (2005) by the American Physical Society. (b) Reprinted with permission from Ref. [26]. Copyright (2006) American Chemical Society. (c) Reprinted with permission from Ref. [32]. Copyright (2011) American Chemical Society. (d) Reprinted from Ref. [35], with permission from the Chinese Academy of Sciences. (e) Reprinted with permission from Ref. [36]. Copyright (2011) by the American Physical Society.

All the SPP excitation methods discussed above use light to generate SPPs. Electrically driven plasmon sources based on different systems have been pursued as well. By using a light-emitting diode (LED), a quantum cascade laser, silicon nanocrystals, a GaAs NW, a GaAs quantum well, and a carbon nanotube, SPPs are electrically excited in metal films, slot waveguides, and nanostrip waveguides [37–44]. As for the NW waveguide, it's demonstrated that voltage-controlled SPPs can be generated by using the tunneling electrons between the gold tip of a scanning tunneling microscope (STM) and a gold NW [36]. As shown in Fig. 6.1e, the locally excited plasmons propagate along the NW and emit at the other end. The NW SPPs can also be excited by high-energy electrons as used in electron energy loss spectroscopy [45, 46]. Different from the optical excitation of NW plasmons, the electronic excitation can be realized at any designated positions on the NW.

6.2.2 SPP Detection

The SPPs are usually detected after they are converted into photons. The detection of SPPs can be regarded as the reverse process of exciting SPPs by photons. In the Kretschmann–Raether configuration, the light from the glass side excites the SPPs at the metal–air interface. In reverse, the SPPs will radiate into the glass medium. For a NW waveguide on a glass substrate and exposed to air, the propagating SPPs can radiate into the glass substrate. This kind of radiation is usually called leakage radiation, while the corresponding plasmon mode is called leaky mode. By using an oil immersion objective with a high numerical aperture (NA), as shown in Fig. 6.2a, the light due to leakage radiation can be experimentally observed. For the NW in Fig. 6.2b, SPPs are launched from the bottom end of the NW by focused laser light. As can be seen, besides the top NW end, the whole NW body is radiating. The radiation intensity is proportional to the local electric field intensity of the leaky plasmon mode; thus the leakage radiation image reveals the local field intensity distribution.

The nanostructure features that can couple light into SPPs can also convert the plasmons into light. Thus the structure roughness or defects on the waveguides can scatter the plasmons into photons. For a metal NW waveguide, the SPPs propagating along the NW cannot be detected in the far field due to the strong localization of the SPP field, while they can be detected by the light scattering at the end of the NW, or other defects on the wire, such as a sharp bending of the wire and the nanoparticles attached to the NW. Figure 6.3 shows three examples, a straight Ag NW, a NW-nanoparticle structure, and a bent NW, with the SPPs excited by focused laser light at one end of the NW and detected as emitted photons at the other end of the NW and the sites with defects (with an attached nanoparticle or a sharp bending) by a charge-coupled device (CCD) detector.

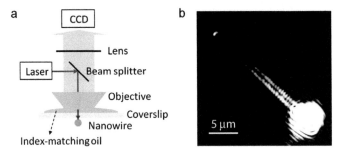

Figure 6.2 (a) Schematic illustration of the experimental setup for leakage radiation imaging. (b) Leakage radiation image of a Ag NW with the laser light focused on the bottom end.

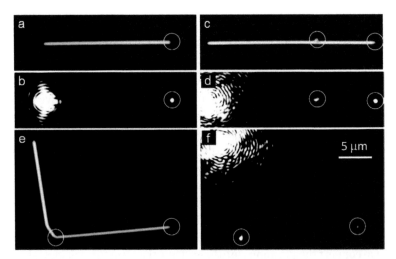

Figure 6.3 (a, c, e) Optical microscopy images of Ag NW structures. (b, d, f) Laser light of 633 nm wavelength focused on one end of the NW excites the propagating SPPs, which couple out as photons at the discontinuous points marked by circles. The scale bar in (f) applies to all images.

By using an SNOM, the electric field distribution of propagating SPPs can be detected with high spatial resolution beyond the diffraction limit. The tapered optical fiber can be used as a near-field probe to detect the propagating SPPs [47]. The tapered fiber probe is usually made by pulling an optical fiber and may be coated by a metal film. The probe is scanned over the sample with a short separation between them and scatters part of the local electric field to the fiber. The collected signal is finally detected by a photodetector to generate a map of the light intensity distribution, which corresponds to the near-field distribution of the propagating SPPs. Figure 6.4 shows the SNOM image for a part of a Ag NW waveguide with the SPPs excited from one end of the NW by focused laser light. The periodic modulation of the near field on the NW originates from the interference of propagating SPPs due to the reflection at the NW end. In the near-field experiments with the plasmons excited by the Kretschmann configuration, the SNOM is usually called a photon scanning tunneling microscope (PSTM) by analogy with an STM measuring the tunneling current between the tip and the sample. Besides the SNOM using a fiber-based near-field probe, a scattering-type apertureless SNOM setup that uses a sharp tip as a near-field probe is also a useful tool for SPP characterization.

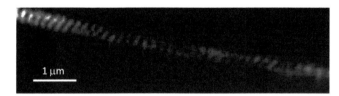

Figure 6.4 The SNOM image of a silver NW coated with 30 nm Al_2O_3. The wavelength of the excitation laser is 633 nm. The radius of the NW is about 150 nm. Excitation light was focused on the NW end using a 100x objective. Reprinted with permission from Ref. [48]. Copyright (2011) American Chemical Society.

As the reverse process of excited nanoemitters launching propagating SPPs, the local plasmon field will excite the nanoemitters near the waveguides when the light wavelength is within the absorption band of the emitters. Therefore, the nanoemitters can function as local probes to reveal the intensity of the plasmon field. If a layer of nanoemitters is coated on the NW surface, by detecting the fluorescence or Raman scattering of the emitters, the local field intensity distribution can be obtained, as the emitting intensity of those probes is proportional to the local electric field intensity. Figure 6.5 shows a silver NW covered by semiconductor QDs. When the laser light is incident on the left end of the NW, the QDs on the NW are excited [49], as shown in the fluorescence image in Fig. 6.5b. To make quantitative analysis of the local field intensity distribution, the coverage of the emitters on the NW should be very uniform. To avoid the fluorescence quenching caused by the metal, some dielectric layer is usually needed as a spacer between the metal and the emitters [48, 49].

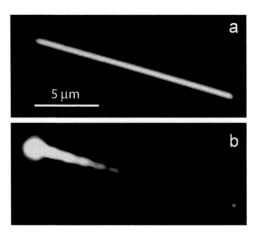

Figure 6.5 (a) Optical microscopy image of a Ag NW. (b) QD fluorescence image with the laser light focused on the left end of the NW. The NW radius is about 150 nm. The Al_2O_3 thickness is 10 nm.

To realize the on-chip plasmonic circuits and the integration of optical and electrical components, the electrical detection of SPPs is essential. By the near-field coupling between a plasmonic NW waveguide and a Ge NW field-effect transistor, electrical SPP detection was demonstrated [50]. As shown in Fig. 6.6a, the Ag NW is crossed over a Ge NW. The SPPs are launched to propagate along the Ag NW and are converted to electron–hole pairs at the cross junction and detected by the current through the Ge NW. The map of current obtained by scanning the focused laser beam over the device shows that besides the direct laser illumination on the Ge NW, the photocurrent is detected when the laser is focused on the two ends of the NW, because the propagating SPPs can only be excited at the ends of the NW. The plasmon-induced photocurrent is strongest when the laser light is polarized parallel to the NW, which provides further evidence for the electrical detection of SPPs. By using single-layer MoS_2, the electrical detection of SPPs in a Ag NW can also be realized, as schematically shown in Fig. 6.6b [51]. Moreover, the electrical detection of SPPs in a metal-insulator-metal waveguide and in a gold strip waveguide was realized by means of an integrated metal-semiconductor-metal photodetector and a quantum point contact, respectively [52, 53].

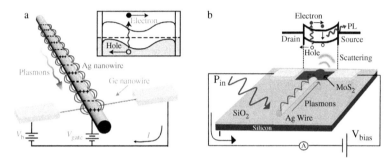

Figure 6.6 Electrical plasmon detection. (a) Schematic diagram of electrical plasmon detector operation. Inset: Electron–hole pair generation and separation in the Ge NW detector. (b) Schematic of the MoS_2/NW detector. Inset: Band diagram showing pathways after electron–hole pair creation. (a) Reprinted by permission from Macmillan Publishers Ltd: [*Nature Physics*] (Ref. [50]), copyright (2009). (b) Reprinted with permission from Ref. [51]. Copyright (2015) American Chemical Society.

6.3 Fundamental Properties of SPPs in Metal Nanowires

6.3.1 SPP Modes in Metal Nanowires

For a metal NW of infinite length embedded in a homogeneous dielectric environment, the surface plasmon wave propagates along the NW with tight spatial confinement of the electromagnetic field. In cylindrical coordinates (r, ϕ, z), the field distributions of the eigenmodes supported by the cylindrical NW can be expressed as [54–56]

$$E_m(r,\phi,z) = A_m R_m(k_{m,\perp} r) e^{im\phi} e^{ik_{m,\|} z}, \tag{6.2}$$

where m is an integer denoting the order of the eigenmodes; A_m is the amplitude of the m-order mode; R_m is the radial distribution of the electric field, which is Bessel functions and Hankel functions of the first kind inside and outside the cylinder, respectively; $k_{m,\perp}$ is the wave vector perpendicular to the NW, describing the decay of the electric field; and $k_{m,\parallel}$ is the wave vector along the NW. Figures 6.7a and 6.7b show the electric field distributions of the two lowest-order modes, that is, $|m| = 0, 1$ for a Ag NW of radius 60 nm in a homogeneous medium of refractive index 1.56 for a vacuum wavelength of 633 nm.

The lowest-order mode has no magnetic field along the NW, that is, it is transverse magnetic (TM) polarized, so it is also called TM_0 mode. As can be seen in Fig. 6.7a, the TM_0 mode is axially symmetric with the electric field radially polarized. For the higher-order modes, they have both electric and magnetic field components along the NW; thus they are hybrid modes. As the magnetic field along the NW is smaller than the electric field along the NW, they are the so-called HE modes. The mode number m characterizes the winding of the mode on the NW. For the m-order mode, there are $2m$ nodes of field on the circumference of the NW cross section. The modes except $m = 0$ are doubly degenerated. The field distributions of the two degenerated modes have an angle shift of $\pi/2m$. For the HE_1 mode shown in Fig. 6.7b, the electric field points to the same direction along x, while to opposite directions along y. Rotating the HE_1 image by 90° results in the field distribution of another degenerate HE_1 mode (these two HE_1 modes will be called HE_1^X and HE_1^Y in the latter).

Figure 6.7c shows the dependences of the effective refractive index n_{eff} and the propagation length L_{spp} of TM_0 and HE_1 modes on the NW radius. For the TM_0 mode, with the decrease of the NW radius, the effective refractive index increases quickly, which means the mode is more tightly confined on the NW surface, resulting in larger Ohmic loss. Therefore, the propagation length of the TM_0 mode is decreased with the decrease of the NW radius. The propagation length here is defined as the distance for the SPP intensity decaying to $1/e$ of the original intensity. For the HE_1 mode, the dependences of n_{eff} and L_{spp} on the NW radius are opposite to the TM_0 mode. With the decrease of the NW radius, n_{eff} of the HE_1 mode approaches the refractive index of the surrounding medium and the electric field cannot be confined around the NW anymore for a very small radius.

For higher-order modes, the effective refractive index decreases fast with the decrease of the NW radius and is cut off at a certain radius. For the metal NW waveguides for applications in plasmonic integrated circuits, the TM_0 and HE_1 modes are most interesting. The excitation of multiple modes makes the SPP propagation in the NW highly controllable.

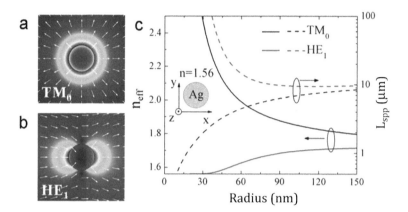

Figure 6.7 Two lowest-order plasmon modes in a silver NW. (a, b) Electric field distribution of TM_0 (a) and HE_1 (b) modes in a Ag NW with a radius of 60 nm. The two modes have different symmetry. (c) The effective refractive index and propagation length of the TM_0 and HE_1 modes as a function of the NW radius. The excitation wavelength is 633 nm. The refractive index of the dielectric environment is $n = 1.56$ [57].

In experiments, the SPPs on a metal NW can be excited by incident light at the NW terminal. Depending on the incident polarization, different plasmon modes can be excited. For excitation by focused laser light at the end of the NW, the excitation can be described by a paraxial Gaussian beam with an instantaneous electric field $\mathbf{E}_{inc}(\mathbf{r}) = \mathbf{E}_0(\mathbf{r})e^{-i\varphi}$, where $\varphi = \omega t$ is the incident phase and $\mathbf{E}_0(\mathbf{r})$ is the mode profile of the incident light [55]. For $\varphi = 0$, the TM_0 and HE_1^X modes will be excited when the incident polarization is parallel and perpendicular to the NW axis, respectively. With a $\pi/2$ phase shift of φ, the modes HE_1^Y and HE_2 will be excited for parallel and perpendicular polarized excitation, respectively. Note that the experimental excitation of the HE_1^Y and HE_2 modes is facilitated by the retardation effects in the optically thick NWs.

Putting a silver NW on substrate dramatically changes the plasmon eigenmodes on the NW since the environment becomes asymmetric. The induced charges on the substrate caused by the oscillating charges of the plasmon modes form an "image" wire. The coupling between the NW and its image will dramatically change the original plasmon modes. Figure 6.8 shows the four lowest-order modes (H_0, H_1, H_2, and H_3) for a silver NW on a glass substrate. As can be seen, the electric field distributions are quite different from that in Fig. 6.7. Considering the symmetry of the electric field, the H_0 mode is analogous to the TM_0 mode, while the H_1 mode is analogous to the HE_1 mode. Input light polarized parallel to the NW can excite H_0 and H_2 modes, and input light polarized perpendicular to the NW can excite H_1 and H_3 modes. The effective refractive indexes of H_1, H_2, and H_3 modes are smaller than the refractive index of the substrate, so they are leaky modes with part of the energy radiating to the substrate. The leakage radiation of these modes enables the imaging of SPPs through leakage radiation microscopy, as shown in Fig. 6.2. The coherent superposition of these modes results in the beating patterns of SPPs on the substrate-supported NWs.

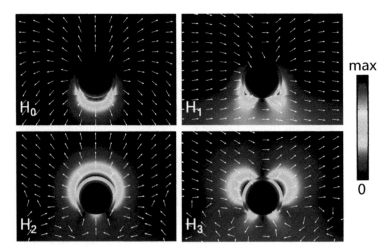

Figure 6.8 The electric field distributions of the four lowest-order modes supported by a Ag NW on a glass substrate. The NW is 160 nm in radius and is coated with an Al_2O_3 layer of 30 nm thickness. Reproduced from Ref. [58] with permission from The Royal Society of Chemistry.

6.3.2 SPP Propagation in Metal Nanowires

The propagating plasmons on metal NWs are the superposition of the eigenmodes discussed before. Figure 6.9 shows the plasmon propagation in a silver NW on a glass substrate. An Al_2O_3 layer of 30 nm thickness was deposited on top of the NW, followed by the spin coating of CdSe@ZnS QDs. The QD fluorescence image by wide-field excitation (Fig. 6.9ii) shows that the distribution of the QDs on the NW is uniform. By launching the propagating SPPs at the top end of the NW (Fig. 6.9iii), the QDs on the NW are excited, showing periodic modulation of the fluorescence intensity (Fig. 6.9iv). As the fluorescence intensity is proportional to the intensity of the electric field, the QD fluorescence image reveals the plasmon field distribution. For parallel polarization, the periodically modulated near-field distribution is due to the superposition of H_0 and H_2 modes. The beating period Λ can be obtained by $\Lambda = 2\pi/\Delta k$, where Δk is the difference of the wave vectors of the two modes. An equivalent form of this equation is $\Lambda = \lambda/\Delta n_{eff}$, where λ is the vacuum wavelength of the excitation light and n_{eff} is the effective refractive index of the mode. Rotating the incident polarization changes the near-field pattern (Fig. 6.9iv–vii). For the polarizations shown in Fig. 6.9v and 6.9vii, the plasmon fields are in zigzag forms and the field distributions are mirror images of each other. These zigzag plasmon fields are caused by the superposition of plasmon modes excited by polarizations along the NW and perpendicular to the NW (mainly H_0 and H_1 modes). When the polarization of the excitation light is perpendicular to the NW (Fig. 6.9vi), the near field is mainly distributed on the two sides of the NW, corresponding to the excitation of the H_1 mode.

The period of the near-field distribution is dependent on the thickness T of the Al_2O_3 layer on the NW. For a NW of radius 155 nm coated with 30 nm of Al_2O_3, the beating period is about 1.7 µm. For thicker Al_2O_3, the period is increased: for $T = 50$ nm, $\Lambda = \sim 2.9$ µm; for $T = 80$ nm, $\Lambda = \sim 5.8$ µm (Fig. 6.10a). For a NW originally coated with 50 nm of Al_2O_3, the period is increased from 2.9 µm to 3.3 µm by depositing additional 5 nm of Al_2O_3 and to 3.8 µm by further adding 5 nm of Al_2O_3 (Fig. 6.10b). The period change is about 90 nm per nanometer of Al_2O_3. The sensitive response of propagating plasmons on the NW to the change of the local dielectric environment can be used for a new type of highly sensitive on-chip optical sensors [59].

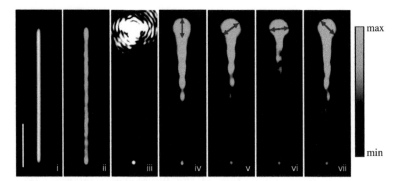

Figure 6.9 QD fluorescence images of plasmons launched by 633 nm laser excitation at one end of a Ag NW. Changing the polarization angle at the input end modifies the field distribution on the NW. (i) Optical image of a Ag NW. (ii) QD fluorescence image with wide-field excitation. (iii) Laser light of 633 nm wavelength focused on the top end of the NW excites the propagating plasmons that couple out as photons at the bottom end of the NW. (iv–vii) QD fluorescence images for different polarizations of incident laser light. The scale bar is 5 μm. The double-headed arrows indicate the laser polarization. Reprinted with permission from Ref. [48]. Copyright (2011) American Chemical Society.

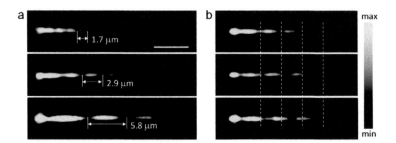

Figure 6.10 (a) QD fluorescence images under excitation at the left ends of NWs. The radius of the NWs is about 155 nm, and the corresponding Al_2O_3 thicknesses are 30, 50, and 80 nm from top to bottom. (b) QD fluorescence images for a 162 nm radius NW with a 50 nm Al_2O_3 coating measured in air (top), then after depositing 5 nm of Al_2O_3 (middle), and finally with an additional 5 nm of Al_2O_3 (bottom). The dashed white lines are visual guides to show the shift of the plasmon near-field pattern [60].

The dependences of the beating period on Al_2O_3 thickness T and NW radius R are shown in Fig. 6.11. The period is increased with the increase of the Al_2O_3 thickness and finally saturated for thick Al_2O_3 (Fig. 6.11a and 6.11c). For a thin coating ($T \leq 50$ nm), the period is increased with the increase of the NW radius and saturated when the NW radius is about 125 nm. For a thick coating, the period is increased monotonically with the increase of the NW radius. The change of the beating period reflects the change of the wave vectors of the plasmon modes according to $\Lambda = 2\pi/\Delta k$. From the calculated wave vectors in Fig. 6.11b and 6.11d, the period can be obtained (lines in Fig. 6.11a and 6.11c), which agrees well with the experimental values.

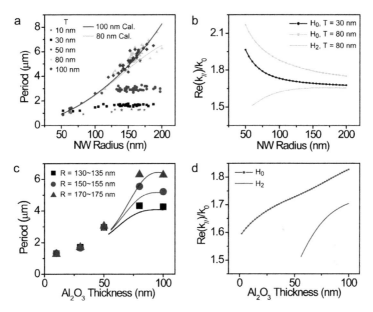

Figure 6.11 (a, b) The beating period (a) and the real part of the wave vector $Re(k_{||})$ of the H_0 and H_2 modes (b) as a function of NW radius. (c, d) The period (c) and $Re(k_{||})$ of the H_0 and H_2 modes (d) as a function of the Al_2O_3 coating thickness. The NW radius is 155 nm for the calculation in (d). In (a) and (c), the dots are experimental data and the lines are calculated data by the finite element method [60].

As the H_2 mode is leaky to the glass substrate, the wave vector of this mode can be experimentally measured using the Fourier imaging method as schematically shown in Fig. 6.12a [61]. The wave vector of surface plasmons on the NW, k_{spp}, and the wave vector of leaky radiation light in the glass substrate, k_{photon}, are related by the phase matching condition

$$\text{Re}(k_{\text{spp}}) = k_{\text{photon}} \sin \theta, \tag{6.3}$$

where θ is the angle between the radiation direction and the normal of the substrate. In experiment, the radiation to the glass side is collected by using an oil immersion objective with a high NA.

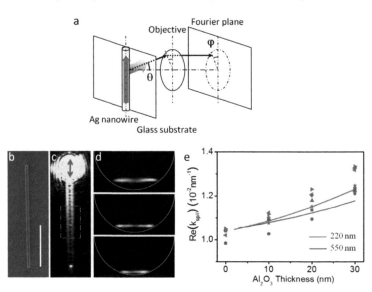

Figure 6.12 Measuring the wave vector of the plasmon mode in experiment. (a) Schematic illustration of the Fourier imaging method. (b) SEM image of a silver NW. The scale bar is 5 µm. (c) Leakage radiation image of propagating plasmons on the silver NW. The dashed rectangle outlines the area for Fourier imaging. (d) Fourier images for different thicknesses of the Al_2O_3 layer. From top to bottom, the Al_2O_3 thickness is 0, 15, and 25 nm, respectively. The green circles on the Fourier images represent the maximum angle θ of radiation that can be collected by the optical system. (e) The wave vector of the H_2 mode for NWs of different diameters and thicknesses of Al_2O_3. The red dots are measured from thick NWs with diameters in the range of 410–640 nm, while the green dots are from thin NWs with diameters in the range of 200–240 nm. The green and red curves are the simulation results for NWs of diameters 220 nm and 550 nm, respectively. Reproduced from Ref. [61], with permission from John Wiley and Sons.

Figures 6.12b and 6.12c show the scanning electron microscopy (SEM) image and leakage radiation image of a silver NW on a glass substrate, respectively. For the Fourier imaging measurement, an aperture was used to block the excitation light and select the area for Fourier imaging, as outlined by the dashed rectangle in Fig. 6.12c.

The top panel in Fig. 6.12d shows the Fourier image of the H_2 mode radiation for the bare Ag NW on a glass substrate, from which the radiation angle and the value of the wave vector for this mode can be obtained. By depositing a layer of Al_2O_3 onto the NW, the radiation angle θ is increased (middle and bottom panels in Fig. 6.12d). The radiation angle θ for Al_2O_3 thicknesses of 0, 15, and 25 nm is 44°, 50°, and 56°, respectively. Figure 6.12e shows the wave vector of the H_2 mode is increased with the increase of the Al_2O_3 thickness and the wave vector for thinner NWs is larger than that for thicker NWs.

The Al_2O_3 thickness also influences the dependence of the beating period on the wavelength. For the Ag NW coated with 30 nm of Al_2O_3, the period is almost the same for excitation by laser light of 532 nm and 633 nm wavelengths (Fig. 6.13a,b), whereas for the NW coated with 50 nm of Al_2O_3, the period for 532 nm excitation is much larger than for 633 nm (Fig. 6.13c,d). These results indicate that the dependence of the beating period on the excitation wavelength can be controlled through the structure geometries. This flexibility is important for keeping signals at different wavelengths together or separating them to follow different paths for different device functions in plasmonic circuits.

The change of the near-field pattern can also be caused by the bulk changes of the surrounding medium. Immersing the glass-supported NW in water or oil rather than in air dramatically changes the near-field pattern, as shown in Fig. 6.14a. When measured in air, the period is about 1.3 µm (top panel). By putting the NW into water, increasing the refractive index of the environmental medium from 1 to 1.33, the period increases to 4.4 µm (middle panel). Replacing the water by oil, which increases the refractive index to 1.51, makes the near-field period increase to 7.2 µm (bottom panel). The change of the period can be understood from the dispersion relations of the SPP modes in the Ag NW. As shown in Fig. 6.14b, compared to the NW in a uniform environment, the difference of the wave vectors, Δk, for the two modes in glass-supported Ag NW is much larger. Therefore, the period of the near-field distribution for the NW on a glass substrate is small according to $\Lambda = 2\pi/\Delta k$. By increasing the refractive index of the surrounding media, the dispersion curves are shifted and the wave vector difference of the two modes is decreased, so the period is increased.

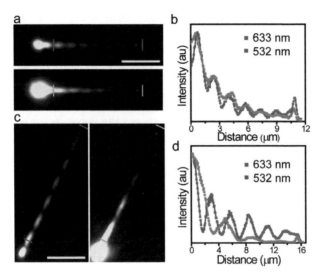

Figure 6.13 Near-field distributions for different excitation wavelengths. (a) QD fluorescence images for a 140 nm radius NW coated with 30 nm of Al_2O_3 for the excitation wavelength of 633 nm (upper) and 532 nm (lower). (b) The intensity profiles along the NW between the short bars in (a). The intensity is normalized by the maximum of the two curves. (c, d) The corresponding data for a 135 nm radius NW coated with 50 nm of Al_2O_3. The scale bars are 5 μm [60].

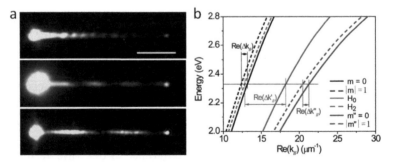

Figure 6.14 (a) QD fluorescence images for a 155 nm radius NW initially coated with 10 nm of Al_2O_3 and QDs and then capped with another 5 nm of Al_2O_3 to protect the water-soluble QDs from being removed, measured in air (top), water (middle), and oil (bottom). The scale bar is 5 μm. (b) Dispersion relations for the $m = 0$ (solid) and $m = 1$ (dash) plasmon modes of a bare Ag NW in air (black) and in a glass matrix (blue) and the H_0 and H_2 modes (red) of a NW supported on an air–glass interface. The radius of the cylindrical NW is 150 nm [60].

When the sample is immersed in oil, the asymmetry of the environment for the glass-supported NW is compensated. The NW is located in a symmetric medium with a refractive index of about 1.51. In a homogenous medium, the SPPs on the NW evolve from zigzag propagation (Fig. 6.9v and 6.9vii) to helical propagation around the NW (Fig. 6.15). Depending on the polarization of the excitation light, the helical plasmons show left-handed or right-handed chirality. The period of the helix is increased with the increase of the NW radius (comparing Fig. 6.15a and 6.15b).

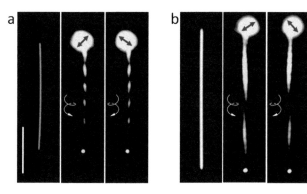

Figure 6.15 Optical images of Ag NWs (left) and QD fluorescence images showing chiral SPPs (middle and right). The white helical arrows highlight the handedness of the plasmon helix. The radius of the NW is 65 nm in (a) and about 150 nm in (b). The scale bar in (a) is 5 μm and applies to (b) as well. The Ag NWs were deposited on glass substrates and coated by QDs with a 10 nm Al_2O_3 layer as a spacer. Linearly polarized laser light of 633 nm was incident on the top end of the NW through an oil immersion objective. (a) Reproduced from Ref. [58] with permission from The Royal Society of Chemistry. (b) Reprinted with permission from Ref. [55]. Copyright (2011) by the American Physical Society.

Figure 6.16 shows the simulation results for the chiral propagation of the SPPs on a Ag NW. When the incident polarization angle is between 0° and 90° (Fig. 6.16a), the TM_0 mode and doubly degenerated HE_1 mode in the NW can be simultaneously excited. As the HE_1^Y mode has a phase delay $\pi/2$ with respect to the HE_1^X mode, coherent interference of these two SPP waves of equal amplitude will result in a circularly polarized guided wave. The simultaneously excited TM_0 mode will interfere with the circularly polarized SPPs, resulting in a spiral near-field pattern. Figure 6.16b shows the time-averaged power flow at different cross sections of the NW, from which the helical propagation of SPPs on the NW can be clearly seen. The phase delay between the two degenerate HE_1 modes determines the handedness of the chiral SPPs. The period of the chiral SPPs is inversely proportional to the difference between the wave vectors of the TM_0 and HE_1 modes, that is, $\Lambda = 2\pi/\Delta k$. When the NW radius is increased, the difference of the two wave vectors is decreased; thus the period is increased (Fig. 6.16c).

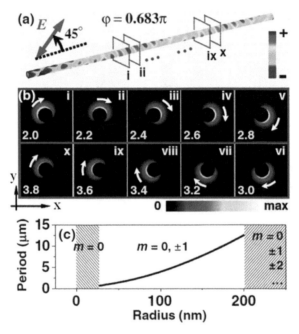

Figure 6.16 (a) Surface charge density plot on a Ag NW. The maximal value of surface charge was truncated at the incident end to better show the plasmon modes on the wire. The length of the simulated NW is L = 5.0 μm and the radius is R = 60 nm. The incident polarization angle is θ = 45° in (a) and (b). (b) Time-averaged power flow in the x-y plane at different positions along the NW, where z = 2.0 to 3.8 μm (i–x) in steps of 0.2 μm, indicated by the blue frames in (a). The white arrows highlight the rotation of electromagnetic energy as a function of the position along the NW. (c) Periods of the plasmon helix as a function of the NW radius. The blue region denotes the single-mode dominant regime, and the magenta region denotes the multimode regime. Reprinted with permission from Ref. [55]. Copyright (2011) by the American Physical Society.

In the above discussions, the excitation of different plasmon modes is controlled by the polarization of the excitation light. The direct conversion of different modes on the NW is highly desired for adjusting the mode proportion and giving flexible control on SPP propagation. It is found that the plasmon modes on the NW can be converted by the symmetry breaking of the structures. Figure 6.17a–c shows the experimentally measured SPP field distributions by QD fluorescence imaging on a Ag NW with a nearby nanoparticle, a branched NW, and a bent NW. In all the QD fluorescence images, the symmetric field distributions become zigzag shapes after the symmetry-broken regions, indicating the generation of new modes. The experimental results are reproduced by simulations, as shown in Fig. 6.17d. The mode conversion occurs due to the redistribution of the electric field on the wave front caused by the scattering of SPPs at the symmetry-broken regions. As branched NWs and bent NWs are fundamental structures to construct functional NW networks, the mode conversion effect is widely present. This mode conversion effect offers a versatile way to control SPP propagation in plasmonic NW networks.

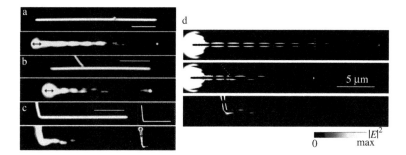

Figure 6.17 SPP mode conversion induced by structural symmetry breaking. (a–c) Experimental observation of mode conversion in three kinds of symmetry-broken NW-based structures: a NW with a nanoparticle placed nearby (a), a branched NW (b), and a bent NW (c). Top: white-light optical images. Bottom: QD fluorescence images. The insets in (c) show the intact images of the bent wire. The double-headed arrows indicate the polarization of the excitation light. The length of the scale bars is 5 µm. The radius of the NWs is about 150 nm. The Al_2O_3 thickness is 50 nm for the structure in (a) and 30 nm for the structures in (b) and (c). (d) Simulation results of electric field intensity distributions for mode conversions in the three kinds of experimental structures. The distribution is on the horizontal section across the axis of the NWs. The radius of all the NWs and the particle is 150 nm. For the nanoparticle-NW and the branched NW, the two component structures are separated by 1 nm [57].

6.3.3 Group Velocity

Since the wave vector of SPPs is larger than that of light in vacuum, that is, the dispersion curve of SPPs is below that of light, the group velocity of SPPs is slower than that of light in vacuum. Allione et al. experimentally determined the group velocity of propagating plasmons in chemically synthesized Ag NWs [25]. In their experiments, Kretschmann configuration was used to excite

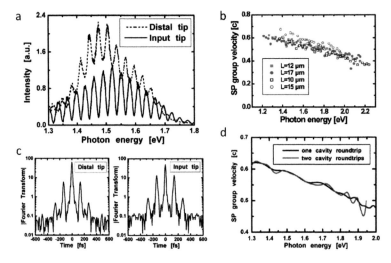

Figure 6.18 (a) Emission spectra from the two tips of a single 12 μm long Ag NW. Both spectra were taken using a focused coherent white-light source for the excitation of the input tip in the Kretschmann configuration. (b) Spectral dependence of SPP group velocity obtained by applying the formula for v_{gr} to the spectra collected with the different excitation sources. Filled symbols: data collected with a spatially coherent broadband excitation source. Empty symbols: data collected under excitation with a halogen lamp. Different symbols correspond to different NWs. (c) Absolute values of the Fourier transforms of the two spectra presented in (a). Note that the horizontal axes in the Fourier transform domain are in time units and the sidebands mimic propagation of ultrashort SPP pulses in the NW cavities. (d) Spectral dependence of SPP group velocity obtained by applying the Fourier-transform-based algorithm to recover the phase of interference fringes presented in (a). The two curves were obtained from the analysis of the first and second sidebands in the Fourier transform domain shown in the left panel of (c), corresponding to single and double roundtrips of surface plasmons in the Ag NW cavity. Reprinted with permission from Ref. [25]. Copyright (2008) American Chemical Society.

the propagating plasmons in the Ag NWs. The spectra of the light collected at both ends of the NW show oscillations caused by the Fabry–Pérot resonances in the NW cavity, as shown in Figure 6.18a. The group velocity of SPPs can be expressed as follows:

$$v_{gr} = 2L\frac{\Delta\omega}{2\pi} = 2Lc\frac{\Delta\lambda}{\lambda^2}, \qquad (6.4)$$

where L is the NW length, c and λ are the light speed and wavelength, respectively, in vacuum, and $\Delta\omega = 2\pi c\Delta\lambda/\lambda^2$ is the spectral fringe spacing in frequency. By analyzing the spectra in Fig. 6.18a using the above equation, the group velocity of propagating plasmons in Ag NWs can be obtained, as shown in Fig. 6.18b. As can be seen, the group velocity is about half of the light group velocity and decreases with the increase of photon energy.

The Fourier transforms of the spectra are shown in Fig. 6.18c. By applying the Fourier-transform-based algorithm [62], the phase of the spectral fringes can be recovered and thus the group velocity can be obtained. Figure 6.18d shows the dependence of the group velocity on the photon energy obtained from the spectrum at the distal tip, which agrees well with the result in Fig. 6.18b. The group velocity of the propagating plasmons in Ag NWs was also measured by far-field spectral interferometry using a femtosecond pulsed laser. It was found that the SPP group velocity is dependent on the NW diameter and decreases drastically as the NW diameter is smaller than 100 nm [63].

6.3.4 Propagation Length and Loss

When SPPs propagate on metal NWs, the intensity will decrease due to the damping caused by the Ohmic loss in the metal and the radiation into the environment. For excitation light of a fixed intensity, the emission intensity at the end of the NW is dependent on the distance between the excitation site and the emission end. This distance-dependent emission intensity can be measured by exciting the SPPs using a nanofiber [32, 64]. As shown in Fig. 6.19a, a tapered fiber is used to launch SPPs by evanescent field coupling. As the excitation site is moved toward the left end of the NW, the output intensity at the right end of the NW is decreased (Fig. 6.19b). The emission intensity for different distances plotted in Fig. 6.19c can be fitted using $I(x) = I_0 e^{-x/L_0}$, where $I(x)$ is the emission intensity after propagating the distance x, I_0 is the input intensity, and L_0 is the propagation length at which the SPP intensity decreases to $1/e$ of the input intensity. From the fitting, the SPP propagation length for this NW is found to be 10.6 μm for the 785 nm wavelength. The propagation loss expressed in decibel (dB), $-10\log(I/I_0)$, increases linearly with the propagation distance (Fig. 6.19c), and the propagation loss per unit length is 0.41 dB/μm. Several factors can influence the SPP propagation length of the NW waveguide, including the wavelength of the excitation light, the NW diameter, the substrate, and the crystal structure and defects of the NW.

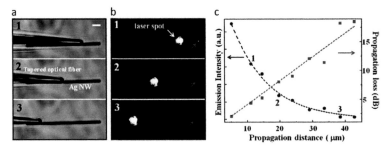

Figure 6.19 Measured propagation losses for varying distances along the Ag NW waveguide. (a) Bright-field optical images and (b) corresponding dark-field optical images are obtained for the Ag NW with a diameter of 750 nm and a length of 45 μm. A 785 nm laser, directed by the tapered optical fiber, is used for the excitation of SPPs in the Ag NW. The light spots of emission at the output end of the Ag NW are shown in (b) 1–3, as indicated by the arrows, which correspond to propagation distances of 11, 19, and 38 μm, respectively. The scale bar is 5 μm. (c) The intensity of the emitted light at the output end of the Ag NW (circles) and the corresponding propagation loss (squares) for different propagation distances. The dotted curve is the exponentially fitted emission intensity for different propagation distances. The dotted line is the linearly fitted propagation loss in dB for varying distances along the Ag NW waveguide. Reprinted with permission from Ref. [32]. Copyright (2011) American Chemical Society.

The SPP propagation length can also be measured by using nanoemitters to excite the SPPs. If the NW is coated by some nanoemitters, such as fluorescent molecules and QDs, the SPPs can be excited by the fluorescent emission. When exciting the emitters at different positions on the NW, the SPP emission intensity at the NW terminal is different depending on the distance between the excitation spot and emitting end of the NW. Figure 6.20a shows the fluorescence images of a molecule-coated Ag NW excited at different positions and the corresponding spectra of the emitted light from the right end of the NW. As can be seen from the spectra, the intensity decreases with the increase in the distance between the excitation spot and the NW end, while the spectra red-shift with the increase in the distance. The redshift of the spectra is due to the difference of the damping rate for different wavelengths. The propagation length of the short-wavelength light is shorter; thus, after long-distance propagation, the red light dominates. From these spectra, the intensity corresponding to different propagation distances for certain wavelengths can be obtained, as shown in Fig. 6.20b. By fitting the distance-dependent emission intensity to the exponential decay function e^{-x/L_0}, the SPP propagation length for the corresponding wavelength can be extracted. Figure 6.20c shows the propagation length L_0 obtained from the fitting is about 3 μm to 6 μm for the wavelength range of 550 nm to 700 nm.

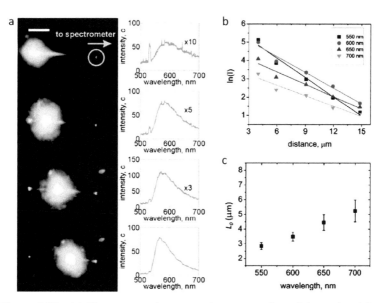

Figure 6.20 (a) Fluorescence images and spectra collected from the right emitting end of a silver NW for different laser spot positions on the wire. The area from which the spectra were collected is outlined with the circle on the image. The white bar in the upper-left corner corresponds to 5 μm length. (b) Logarithm of fluorescence intensity as a function of the distance between the excitation spot and the emitting wire end and corresponding linear fits. (c) Plasmon propagation length (L_0) for 550, 600, 650, and 700 nm wavelengths extracted from the fits. Reprinted from Ref. [65], with the permission of AIP Publishing.

If the NW is bent, additional loss will be introduced at the curved part of the NW. By using a tapered tip, the NW can be bent to different curvatures. By measuring the output intensity of the NW under different bending radii (Fig. 6.21a,b), the bending loss in the NW waveguide can be obtained. When the bending radius is decreased, the output intensity at the NW end is decreased. For a curved NW, both the propagation loss and the bending loss contribute to the intensity decrease at the output end. The output intensity can be expressed as

$$I = I_0 e^{-l/L_0} e^{-\alpha(l-x_0)}, \tag{6.5}$$

where I_0 is the input intensity, α is the parameter to depict the energy attenuation per unit length around the bending section due to bending loss, l is the length of the NW, and x_0 is the length of the straight segment as shown in the inset of Fig. 6.21c. The first exponential decay component e^{-l/L_0} depicts the propagation loss for the straight NW, and the second exponential decay component $e^{-\alpha(l-x_0)}$ depicts the loss caused by the bend of the NW, that is, bending loss. The bending loss ρ can be expressed in dB as

$$\rho = -10\log e^{-\alpha(l-x_0)} = -10\log\frac{I}{I_0} - \frac{10}{2.3}\left(\frac{l}{L_0}\right). \tag{6.6}$$

The experimentally measured bending loss is plotted in Fig. 6.21c as a function of the bending radius R. The bending loss data can be fitted mathematically if the attenuation coefficient takes the form $\alpha \approx C_g e^{-\beta R}$, where C_g and β are fitting parameters depending on the geometries and properties of the NWs. The bending loss is strongly dependent on the bending radius. As the bending radius is decreased further, cracks may happen to the NW. At these crack sites, plasmons can be strongly scattered and thus the output intensity at the NW end is significantly decreased. In addition, the energy loss caused by the NW bending is also dependent on the wavelength of the excitation light and the diameter of the NW.

The loss of the NW SPPs depends strongly on the permittivity of the substrate. A substrate with a larger refractive index leads to higher SPP loss due to the electromagnetic coupling between the surface charges on the NW and the induced charges at the substrate surface. Putting a silver NW on a Si substrate will strongly damp the propagating SPPs. To decrease the SPP loss for Si-supported NWs, low-refractive-index materials like SiO_2 can be deposited on top of Si. A SiO_2 layer of about 200 nm thickness can significantly reduce the Si-caused SPP loss [66]. The leakage radiation of SPPs to the substrate is another channel for SPP damping. To increase the propagation length of SPPs on metal NWs, a scheme using a layered substrate to block the leakage radiation of SPPs is theoretically proposed [67]. The metal NW is placed on a low-permittivity substrate (SiO_2) coated with a high-permittivity layer (Si). The added Si layer acts as an optical barrier to block the leakage radiation to the SiO_2 substrate, so the propagation loss of the plasmons on the NW is reduced. This configuration provides a simple way to improve the performance of the metal NW waveguide. Moreover, using Si as the optical barrier layer also makes this scheme compatible with current silicon technologies. By further adding a SiO_2 layer on top of the Si layer in the above structure, the electric field can be squeezed into the SiO_2 nanogap between the Ag NW and the Si layer, which improves the SPP guiding properties [68].

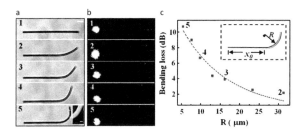

Figure 6.21 Measured bending loss for varying bending radii. (a) Images 1–5 are the selected bright-field optical images demonstrating the bending process of the Ag NW with a diameter of 750 nm, a length of 45 μm, and bending radii of ∞, 32, 16, 9, and 5 μm, respectively. The inset is the SEM image of the curved section with a scale bar of 350 nm. The scale bar in (a) is 5 μm. (b) Images 1–5 are the dark-field optical images for the corresponding bending process. The light spots of emission at the bent end of the Ag NW are indicated by the arrows, and a 785 nm laser is used for the excitation. (c) Pure bending loss as a function of the bending radius. The curved line is the exponential fit to the bending loss in the Ag NW. Inset: geometry of the bent wire. The length of the straight section is x_0, and the bending radius is R. Reprinted with permission from Ref. [32]. Copyright (2011) American Chemical Society.

6.3.5 Emission Direction and Polarization

The propagating SPPs on the NW can emit as photons at the end of the NW, as the reverse process of SPP excitation by light illumination at the NW end. Figure 6.22a shows the emission characteristics of the TM_0 and HE_1^x modes at the end of the NW in a homogeneous dielectric environment [56]. As can be seen, the emission fields maintain the polarization symmetry of the propagating modes on the NW. The emission of the TM_0 mode is radially polarized with axial symmetry, while the emission of the HE_1 mode is linearly polarized. For the TM_0 mode, the distribution of emission is angular and the maximum intensity is along the direction of about 30° with respect to the axis of the NW. For the HE_1 mode, the maximum intensity is along the axial direction, with a small spreading angle of about 15°. The emission features of the plasmon modes can be intuitively understood as results of the dipole emission at the end of the NW.

The SPP emission at the NW end is the superposition of the emissions from different plasmon modes. By illuminating the NW end with laser light, both TM_0 and HE_1 modes can be excited. The excitation of two plasmon modes leads to asymmetric field distribution on the NW and asymmetric emission at the NW end, as shown in the middle panel of Fig. 6.22b. The emission intensity peaks at $\theta \approx 60°$, as can be seen in the bottom panel of Fig. 6.22b. Depending on the NW dimensions and geometries, the maximum emission intensity in the θ direction is in the range of about 45°–60° [69].

A direct way to experimentally measure the emission direction is to image the intensity distribution on the back focal plane (Fourier plane) of the objective [61, 70, 71], as illustrated in Fig. 6.12a. The Ag NW on the glass slide is in a quite asymmetric dielectric environment. In such an asymmetric surrounding, the propagating SPPs can radiate into the substrate if the effective refractive index of the plasmon mode is smaller than the refractive index of the substrate. Therefore, besides the NW end, the NW trunk also emits light. Figure 6.23a shows the leaky radiation image of the H_2 mode excited by laser light polarized parallel to the NW. An aperture was used to select the radiation from a specific area for Fourier imaging.

For the emission from the NW trunk, it shows a straight line shape in the Fourier image, with the emission angle determined by the phase matching condition of SPPs and photons. The emission angle along the NW direction is about 44° (Fig. 6.23b). The emission from the NW end shows an arc-shaped pattern on the Fourier plane (Fig. 6.23c), which is different from the emission of the NW trunk. The radiation angle from the NW end is about 41° for this NW.

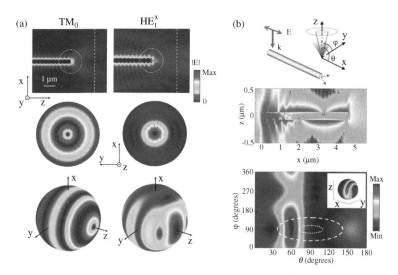

Figure 6.22 (a) Emission characteristics of TM_0 and HE_1^x modes for a Ag NW in a homogeneous dielectric environment. Top row: electric field amplitudes on the cut plane across the axis of the wire; middle row: electric field on the cut plane 2 μm away from the output terminal marked by the dashed line in the top row; bottom row: far-field angular emission distributions transformed from the spheres marked by the dot circles in the top row. The NW radius is 200 nm. The excitation wavelength is 633 nm. The refractive indexes of the surrounding medium and silver are 1.56 and $0.0562 + 4.2776i$, respectively. (b) Top: the coordinates; middle: the calculated distribution of the Poynting intensity around the NW; bottom: the calculated emission intensity as a function of angles φ and θ obtained by far-field transformation. The inset shows the corresponding angular distribution on the integration sphere. The white rings of different sizes are the emission angles that can be collected by the objective for NA = 1.2 and 0.5. The diameter of the NW is 158 nm, and the length is 4.6 μm. The excitation light of 633 nm wavelength is incident on the NW end and polarized along the wire. (a) Reprinted from Ref. [56], with permission from the Chinese Academy of Sciences. (b) Reprinted with permission from Ref. [69]. Copyright (2009) American Chemical Society.

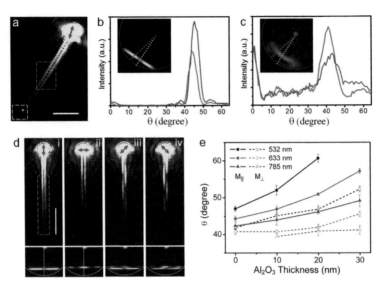

Figure 6.23 (a) Leakage radiation image of propagating plasmons on a silver NW of radius ~150 nm. The wavelength of the excitation light is 633 nm. The scale bar is 5 μm. (b, c) Intensity distribution versus θ along two directions $\varphi = 0$ (green curves) and $\varphi = -15°$ (red curves) on the Fourier images, as marked by dashed green and red lines in the insets. The Fourier images are obtained from the areas marked by dashed rectangles in (a), corresponding to the radiation from the NW trunk and terminal, respectively. (d) Leakage radiation images and corresponding Fourier images for different excitation polarizations for a silver NW of radius ~180 nm. The wavelength of the excitation light is 532 nm. The scale bar is 5 μm. (e) The radiation angles of two modes as a function of Al_2O_3 thickness for excitation wavelengths of 532, 633, and 785 nm. The solid symbols and hollow symbols are experimental means of five NWs for the longitudinal mode and the transverse mode, respectively. The error bars show the standard deviation. The lines are used as a guide to the eye. The double-headed arrows indicate the laser polarization. The green circles on the Fourier images represent the maximum angle θ of the radiation that can be collected by the optical system. (a, b, c) Reproduced from Ref. [61], with permission from John Wiley and Sons. (d, e) Reproduced from Ref. [72] with permission from The Royal Society of Chemistry.

For laser polarization perpendicular to the NW, the H_3 mode can be excited. Figures 6.23d–i and 6.23d–ii show the leaky radiation images (top panels) and Fourier images (bottom panels) for parallel and perpendicular polarizations, respectively. The Fourier images show that the radiation angle for the H_3 mode (excited by perpendicular polarization) is smaller than that for the H_2 mode (excited by parallel polarization). This is because the wave vector of the H_2 mode is larger. For polarization angles of 45° and −45° with respect to the NW, both leaky modes are excited. The coherent superposition of these two modes leads to the zigzag distribution of the leaky radiation (Fig. 6.23d–iii,iv). Figure 6.23e shows the dependences of the radiation angle for the two modes on the wavelength and Al_2O_3 thickness. Since a shorter wavelength and a thicker Al_2O_3 thickness correspond to a larger wave vector of surface plasmons, the corresponding radiation angle is larger according to Eq. 6.3.

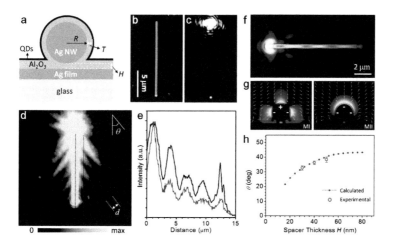

Figure 6.24 (a) Schematic cross section of the sample. (b) SEM image of a Ag NW/Al_2O_3/Ag film sample with R = 200 nm, H = 40 nm, and T = 30 nm. (c) Optical image of the sample with laser light of 633 nm wavelength focused on the top end of the NW. (d) QD fluorescence from the sample that is excited by laser light focused on the top end of the NW and polarized parallel to the NW. (e) Intensity profiles of QD fluorescence measured along the marker lines in (d). The two curves are normalized with the maximum intensity. (f) Top view of the simulated electric field |**E**| distribution at the surface of the top Al_2O_3 layer. (g) Simulated electric field distributions of the two leaky SPP modes on the NW. Surface charges are schematically drawn. The sample parameters in the simulations are R = 200 nm, H = 40 nm, T = 30 nm, and Ag film thickness = 120 nm. The wavelength is 632.8 nm. (h) Calculated and experimental values of the propagation angles of collimated SPP beams as a function of Al_2O_3 thickness. Reprinted with permission from Ref. [73]. Copyright (2015) American Chemical Society.

It is noted that the leakage radiation of SPPs is determined by the phase matching condition so that it is not limited to a glass substrate.

In the composite structure of Ag NW–Al$_2$O$_3$-Ag film (Fig. 6.24a), the SPPs on the Ag NW can radiate to the Ag film to generate SPP waves propagating on the film surface along the directions decided by Re($k_{\text{spp-NW}}$) = $k_{\text{spp-film}}$ cos θ, where θ is the angle between the NW and the propagation direction of SPPs on the silver film. For laser polarization parallel to the NW, two plasmon modes on the NW radiate to the Ag film (Fig. 6.24g), generating two SPP waves on the film propagating along two different directions θ_{I} and θ_{II}. The interference of the two SPP waves produces periodic collimated SPP beams along the direction $\theta = (\theta_{\text{I}} + \theta_{\text{II}})/2$, as shown in Fig. 6.24d. The simulated electric field distribution in Fig. 6.24f reproduces the experimental result. The direction angle θ of the collimated beams increases with the increase of the dielectric spacer thickness (Fig. 6.24h). The directional launching of SPPs on the film by the SPPs propagating on the NW can be regarded as a planar analogue of Cherenkov radiation.

The polarization of the emitted light is also determined by the SPP modes on the NW. As discussed in Section 6.3.1, for a perfect thin cylinder, different plasmon modes can be excited by incident light of different polarizations (Fig. 6.25a). The shape of the NW end can strongly influence the excitation and emission of different plasmon modes. Thus the emission polarization strongly depends on the specific nanostructure of the NW end. For a NW with flat ends (Fig. 6.25b), the excitation of SPPs by parallel incident polarization will result in the emission polarized along the NW (Fig. 6.25c), when detecting the emission in the way as in experiments using an objective. However, for a NW with a side-cut end for excitation or for emission, the polarization of the emitted light is rotated away from the direction along the NW for parallel polarized incident light (Fig. 6.25d–g) [74]. If the morphology of the NW ends can be controlled precisely, it's possible to selectively realize polarization-maintaining or polarization-rotating functions in different applications.

As we have discussed, the SPPs propagate helically along the metal NW in a uniform medium. As a result, the emitted photons at the NW terminal may preserve the chirality of the chiral SPPs. To study the polarization characteristic of the emitted photons, the degree of circular polarization C is defined as

$$C = \frac{2\langle E_x(t)E_y(t)\sin(\delta_x - \delta_y)\rangle}{\langle E_x^2(t)\rangle + \langle E_y^2(t)\rangle + \langle E_z^2(t)\rangle}, \qquad (6.7)$$

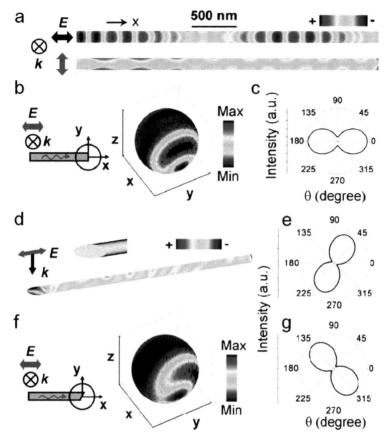

Figure 6.25 (a) Charge distribution on the surface of a Ag NW with flat ends under excitation polarized parallel and perpendicular to the wire. (b) The spatial distribution of the emission intensity on a sphere enclosing the wire end excited by the parallel polarization. (c) Emission intensity of different polarizations from the wire under parallel polarized excitation. (d) A 3D view of the charge distribution on the wire surface under the excitation of parallel polarization. The wire has a 25° side-cut incident end and a flat emission end, as shown in the inset. (e) Polarization of the emission from the wire in (d) under parallel polarized excitation. (f) Spatial distribution of the emission intensity on a sphere enclosing the wire end under the parallel polarized excitation. The cylindrical wire has a flat incident end and a 60° side-cut emission end. (g) Polarization of the emission from the wire in (f) under parallel polarized excitation. The wavelength of the excitation light is 633 nm. The length and diameter of the NW are 3.36 μm and 130 nm, respectively. Reprinted with permission from Ref. [74]. Copyright (2010) American Chemical Society.

where $\langle\ \rangle$ denotes time average and $\delta_x - \delta_y$ is the phase difference between the two transverse electric field components E_x and E_y. The value of C is calculated in a vertical plane 200 nm beyond the output end of the NW. From the calculated spatial map of C shown in Fig. 6.26a, it can be seen that a degree of circular polarization as high as 0.9 can be obtained. The figure of merit (FoM) is defined as $f = I \times C^2$. $I = |\mathbf{E}(\mathbf{r})|^2/|\mathbf{E}_0(0)|^2$ is the relative intensity, where $\mathbf{E}_0(0)$ is the incident electric field at the origin. The map of the FoM shown in Fig. 6.26b confirms the high degree of circular polarization of the optical field at the emission end of the NW. The polarization state of the emission depends strongly on the polarization angle of the linearly polarized incident light. As shown in Fig. 6.26c, C gets to the maximum when the polarization angle θ is 45° and 135° ($\theta = 0$ is along the NW). Since the metal NW can support a continuum of SPPs, the emitted light over a wide spectral range can maintain a high C, that is, circularly/elliptically polarized. This makes metal NWs a good candidate for broadband nanosource of circularly polarized

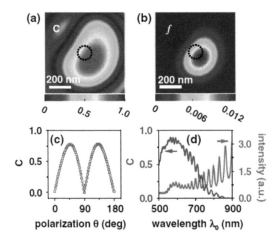

Figure 6.26 (a, b) Calculated maps of degree of circular polarization C (a) and figure of merit f (b) in a vertical plane 200 nm beyond the distal end of a Ag NW. The radius of the NW is 60 nm, and the length is 5.0 µm. The wavelength of the incident laser light is 632.8 nm, and the polarization angle is 45° relative to the NW. The dotted black circles indicate the cross section of the NW. (c, d) C at the center of (a) on the symmetric axis of the NW as a function of θ (c) and vacuum wavelength λ_0 (d). The transmission spectrum (green) is also shown in (d). Reprinted with permission from Ref. [55]. Copyright (2011) by the American Physical Society.

light for nanophotonic applications. For example, it is possible to use the metal NW light source as a new type of probe for scanning near-field optical microscopy or tip-enhanced Raman spectroscopy to study the interaction between circularly polarized light and chiral molecules or nanostructures.

6.3.6 Spin–Orbit Interaction of Light in Plasmonic Nanowires

When circularly polarized light interacts with a medium, the conservation of optical angular momentum can lead to the coupling between the photons' intrinsic spin degree of freedom and their motion. This spin–orbit interaction (SOI) of light is usually very weak, but it can be greatly enhanced in plasmonic nanostructures. Maxwell's equations can be written in a Dirac-like form [75, 76]:

$$c(\hat{\boldsymbol{\alpha}}\cdot\hat{\mathbf{p}})\Phi + \hat{\boldsymbol{\beta}} V\Phi = i\partial/\partial t\,\Phi, \tag{6.8}$$

where c is the speed of light in vacuum; Φ is the field wave function formed by electric and magnetic fields $\Phi = \sqrt{(4\omega)^{-1}}\left[\sqrt{\varepsilon_0}\mathbf{E}\quad \sqrt{\mu_0}\mathbf{H}\right]^T$, where ω is the frequency of light; $\hat{\boldsymbol{\alpha}}_i$ (i = 1, 2, 3) and $\hat{\boldsymbol{\beta}}$ are the four Dirac matrices, among which $\hat{\boldsymbol{\alpha}}_i$ are expressed using the spin-1 Pauli operators; $\hat{\mathbf{p}}$ is the ordinary momentum operator $-i\nabla$; and V is the optical potential induced by the dielectric medium, which is $\omega[1 - \varepsilon(\mathbf{r})]$ for nonmagnetic materials. For light of certain frequency, the optical potential is exclusively determined by the specific magnitude and spatial distribution of the optical constant of the material. The nonmagnetic metal nanosphere can be viewed as a spherical potential well, with the V inside the nanosphere as $\omega(1 - \varepsilon_m)$ and outside as $\omega(1 - \varepsilon_d)$, where ε_m and ε_d are the permittivity of the metal and dielectric material, respectively. The spatial variance of optical material leads to the SOI of photons. Especially at the interface between metal and dielectric material, the large discrepancy of optical constant will result in a large gradient of optical potential and a strong SOI.

Using the above formalism, the orbital momentum density along the phase gradient is defined by $\langle\Phi|-i\nabla|\Phi\rangle$, giving

$$\mathbf{p}^o = (4\omega)^{-1}\,\mathrm{Im}[\varepsilon_0\mathbf{E}^*(\nabla)\mathbf{E} + \mu_0\mathbf{H}^*(\nabla)\mathbf{H}]. \tag{6.9}$$

The orbital momentum density $\mathbf{p}^o$ corresponds to the photon trajectory. The streamlines $l(\mathbf{r})$ of $\mathbf{p}^o$ are traced on the basis of its direction, fulfilling $\dot{l}(\mathbf{r}) \times \mathbf{p}^o = 0$. Figure 6.27a shows the streamlines of $\mathbf{p}^o$ for a gold nanosphere with a radius of 80 nm illuminated by a circularly polarized plane wave of 785 nm wavelength. The orbital momentum density $\mathbf{p}^o$ of the incident light is uniformly forward along the wave vector in the z direction. In the near field of the nanoparticle, it gains azimuthal components and forms a vortex, which manifests as the streamlines of the $\mathbf{p}^o$ twisting around the nanosphere, with the chirality determined by the spin direction of the incident photons. The drastically twisted chiral streamlines reveal that the photons' SOI is strongly enhanced in the surface plasmon near field. Moreover, the circular polarization degree of the field near the gold surface is low, implying most of the spin angular momentum of the incident photons is converted to the orbital angular momentum by means of the SOI.

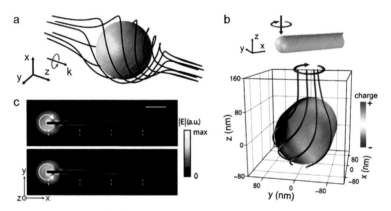

Figure 6.27 (a) A strong SOI in the scattering of circularly polarized photons on a gold nanosphere. The twisted curves are calculation results for streamlines of the orbital momentum density $\mathbf{p}^o$ during the scattering. (b) The SOI on a gold NW. Top: Schematic illustration for the excitation of SPPs on a NW using circularly polarized photons. Bottom: Distributions of surface charges and the streamlines of $\mathbf{p}^o$ for the near field on the tip of a gold NW. (c) Sectional view of the electric field distribution on the x-y plane across the center of the NW under excitation with opposite circular polarizations denoted by the circular arrows. The scale bar is 1 µm. The white dashed lines mark the nodes in the periodic zigzag patterns. The radii of the gold nanosphere and the gold NW are both 80 nm. The refractive index of the surrounding dielectric medium is 1.5. The wavelength is 785 nm. Reprinted with permission from Ref. [76]. Copyright (2016) by the American Physical Society.

The enhanced photonic SOI can also occur in metal NWs. By illuminating the end of a gold NW using a circularly polarized Gaussian beam, twisted orbital momentum density flow and transverse propagation of surface charge wave are observed, indicating the strong SOI process on the NW end (Fig. 6.27b). The transverse components of the orbital momentum density in the *x-y* plane lead to directional coupling of propagating SPPs on the NW, resulting in the periodic zigzag pattern of the electric field distribution (top panel in Fig. 6.27c). The field near the tip is concentrated on one side of the NW and forms the first node of the propagating SPPs. For an excitation beam with opposite circular polarization, the field distribution will be reversed as the mirror image (bottom panel in Fig. 6.27c). Therefore, the incident photons with different spin angular momentums are separated into different spatial trajectories due to the SOI of light on the NW tip, similar to the photonic spin Hall effect.

6.3.7 Nanowire–Emitter Coupling

The NW plasmons can interact with the nearby quantum emitters, such as molecules and QDs. The processes existing in the NW-emitter system are summarized in Fig. 6.28a. The incident light excites the SPPs propagating along the NW, which can be detected by the photons coupled out at another end of the NW (process I). The propagating plasmons in the NW can excite the emitters, QD, for example, near the NW (process II). If an excited emitter is close to a plasmonic NW, the exciton decay will generate surface plasmons in the NW, which propagate along the NW and finally couple out at the NW ends (process III). The processes II and III are shown by the experimental results in the middle and bottom panels of Fig. 6.28b, respectively. In the middle panel, green laser light was focused onto the end C of the NW to launch the propagating SPPs, while in the bottom panel, the green laser light was focused onto the QD A to make the QD excited. Figure 6.28c shows the time traces of fluorescence counts from QD A and NW ends B and C for the laser excitation at the NW end C. As can be seen, the emission at the three spots shows a synchronous blinking behavior, indicating only one QD is coupled with the NW. Actually, the results in the middle panel of Fig. 6.28b and in Fig. 6.28c are the products of combined processes II and III. QD A excited by the propagating SPPs on the NW (process II) decays partly by generating the NW SPPs (process III), with frequencies corresponding to the QD emission, so light emission from NW ends B and C is observed. The plasmon–exciton coupling enables the remote excitation of the QD on the NW and the remote detection of the QD emission. The interaction of plasmons and excitons can be utilized to combine metal nanostructures and semiconductor or organic structures for developing active nanophotonic devices.

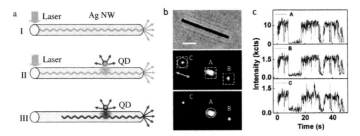

Figure 6.28 (a) Schematic illustration of the processes in the NW-QD system. The wavy lines on the NW represent the propagating SPPs. (b) Top: optical transmission image of a silver NW. The scale bar is 2 μm. Middle: fluorescence image showing the remote excitation of a single QD by propagating SPPs on the Ag NW, corresponding to II in (a). The circle shows the laser excitation position. The double-headed arrow shows the polarization of the excitation light. The largest bright spot A corresponds to the fluorescence from the QD, while two smaller spots B and C correspond to SPPs scattered from the two ends of the NW. Bottom: fluorescence image for excitation light on QD A, corresponding to III in (a). (c) Time traces of fluorescence counts of QD A and scattered light at the NW ends B and C, corresponding to the middle panel in (b). The intensity unit kcts means 1000 counts. The dashed boxes in (b) show the regions where the counts of each pixel are integrated to generate the emission intensity. (a) Reprinted with permission from Ref. [49]. Copyright (2009) American Chemical Society. (b, c) Reprinted from Ref. [35], with permission from the Chinese Academy of Sciences.

The remote excitation technique based on the propagating SPPs can also be used for surface-enhanced Raman scattering (SERS). For a coupled structure of a Ag NW and a Ag nanoparticle (Fig. 6.29a), the electromagnetic coupling between the NW and the nanoparticle results in a greatly enhanced local electric field in the nanogap between them. The nanogap is the hot spot for SERS [77].

If the molecules are located in the nanogap, the Raman scattering signal will be largely enhanced and become detectable. As can be seen in Fig. 6.29a, when the laser was focused on the left end of the NW, only the molecules in the NW–nanoparticle junction were detected, that is, the molecules at the NW–nanoparticle junction were remotely excited [78]. If there are more nanogaps created by attached nanoparticles or NWs on the NW waveguide, the Raman scattering at these nanogaps can be simultaneously excited by the remote excitation way, as shown in Fig. 6.29b. Therefore, multisite remote excitation SERS sensing can be achieved. Compared with conventional excitation configurations, the remote excitation SERS has distinct advantages. Since the enhanced local field in the nanogap is produced by the propagating plasmons, for the molecules in the nanogap, the excitation is background free. The size of the nanogap determines that the size of the excitation source is nanometer scale. This nanoscale source avoids the unnecessary and even harmful large-area illumination to the sample that may cause damage to the sample and produce a high background signal. The sensitivity of remote SERS is very high, so even single-molecule SERS can be detected using this remote excitation technique.

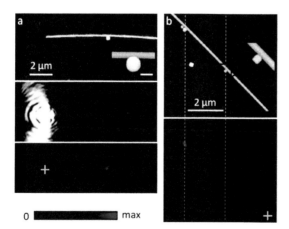

Figure 6.29 Remote excitation SERS of malachite green isothiocyanate (MGITC) molecules excited through propagating plasmons. (a) Top: SEM images of a NW-nanoparticle system; middle: laser light scattering image; bottom: Raman image. (b) Remote SERS in a structure with multiple hot spots. Top: SEM images; bottom: Raman image. The cross marks the position of the laser spot. Reprinted with permission from Ref. [78]. Copyright (2009) American Chemical Society.

The plasmon–exciton interaction in the coupled system of plasmonic waveguide and quantum emitter provides a new approach to manipulate the interaction of single photons/plasmons and quantum emitters, which is important for quantum optics and quantum information sciences. Therefore, the NW–emitter coupled system has drawn the attention of many researchers. Chang et al. theoretically studied the coherent coupling between individual optical emitters and plasmonic waveguides, as well as nanotips [79]. They proposed a scheme for a single-photon transistor using the strong coupling between the propagating plasmons on a metal NW and individual optical emitters [80]. Akimov et al. experimentally studied the generation of single plasmons through the excitation of a QD on a Ag NW [81]. Kolesov et al. showed that single surface plasmons excited by single nitrogen vacancy (NV) defects in diamond nanocrystals exhibit both wave and particle properties, similar to single photons [82]. Liu et al. investigated the coupling between a pair of gold NWs and a nanosized emitter by using the finite-difference time-domain (FDTD) method, and their results show that the converted energy from the emitter to propagating SPPs is increased compared to a single NW [83]. Huck et al. employed an atomic force microscope to manipulate the nanodiamonds with single NV centers to control the coupling between a Ag NW and a single NV center [84]. Kumar et al. studied the coupling of a single NV center with two Ag NWs and the routing of single plasmons in two wires [85, 86]. In a system of two quantum emitters positioned near a plasmonic waveguide, the entanglement of two qubits mediated by the plasmonic waveguide was theoretically investigated by several groups [87–89]. Li et al. experimentally studied the coupling of two QDs with a Ag NW by using a super-resolution optical imaging method to resolve two QDs within the diffraction-limited area [90]. In the coupled system of a single QD and a Ag NW, the decay rates of all the exciton recombination channels, that is, direct free-space radiation channel, surface plasmon generation channel, and nonradiative damping channel, were experimentally determined, so the quantum yield of single plasmons was obtained [91].

6.4 Plasmonic Devices and Circuits

6.4.1 SPP Router, Splitter, Demultiplexer, Switch, and Spin Sorter

To build a plasmonic circuit, splitting and routing plasmon signals to different paths are desired. The mechanism of the signal routing can be examined in a simple system composed of a single NW and a single nanoparticle (Fig. 6.30i). The SPPs launched by the incident light at the top end of the NW can couple out at both the bottom end and the nanoparticle position. The output intensity at these two points depends strongly on the polarization of the excitation light. Stronger emission is observed at the particle position for the incident polarization shown in Fig. 6.30iii, while almost no light is emitted at the particle position for the polarization in Fig. 6.30v. This polarization dependence of the emission intensity indicates that the polarization of the input light can be used to control the signal distribution in different routes in a NW network. By using the QD fluorescence imaging technique, the electric field distribution along the NW was imaged (Fig. 6.30iv,vi). For the polarization in Fig. 6.30iii, the QD image shows that the NW–nanoparticle junction is overlapped with one of the antinodes in the plasmon field distribution pattern on the NW and thus the local electric field intensity at the junction is stronger. The strong near field determines the strong photon emission intensity. For the polarization in Fig. 6.30v, the plasmon field is distributed at the two sides of the NW. The stronger local field intensity is distributed at the opposite side of the NW with respect to the junction, so the photon emission intensity at the particle position is quite low. These results show that the control to the near-field distribution determines the control to the plasmon-routing behavior.

If the nanoparticle is extended to a NW, a branched structure composed of two NWs is formed. The structures composed of multiple NWs can be assembled by using a micromanipulator. The plasmons generated in the primary NW can be directed to different ends of the branched NW structure by controlling the incident polarization [92]. Figure 6.31a shows the propagation of SPPs in a branched Ag NW structure. The intensity at the two output terminals varies with the rotation of the incident polarization. The SPP propagation to the different ends of the structure is wavelength dependent, as shown in Fig. 6.31b–d. When two laser beams of 633 nm and 785 nm wavelengths are input into the nanobranch, the 633 nm light mainly propagates to terminal 2, while the 785 nm light mainly propagates to terminal 3. Thus this simple branched NW structure can function as a plasmon router and demultiplexer.

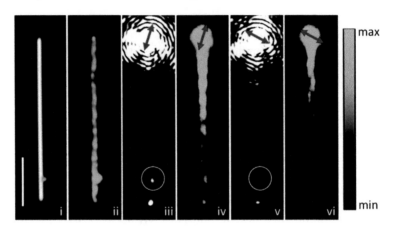

Figure 6.30 Changing the polarization angle at the input end of the NW controls the emission from an adjacent Ag nanoparticle. (i) Optical image of a NW-nanoparticle system. (ii) QD fluorescence image with wide-field excitation. (iii) Scattering image. (iv) QD fluorescence image corresponding to (iii). (v, vi) Scattering and QD fluorescence images for a different polarization. The scale bar is 5 µm. The double-headed arrows indicate the laser polarization. Reprinted with permission from Ref. [48]. Copyright (2011) American Chemical Society.

The wavelength response is clearly demonstrated by exciting the SPPs using a white-light source. Figure 6.32a shows a structure composed of three Ag NWs excited by supercontinuum light. The emission at the NW ends A and B shows clear wavelength dependence. The spectral response can be changed by adding a thin layer of Al_2O_3. As shown in Fig. 6.32b, by adding 10 nm of Al_2O_3 (blue curves in Fig. 6.32b), the NW end A mainly emits light of short wavelengths around 630 nm, while the end B mainly emits light of long wavelengths around 800 nm. The QD fluorescence images excited by laser light of 633 nm in Fig. 6.32c show how the Al_2O_3 thickness influences the plasmon routing in the NW network. For the original structure, the right branch junction is at a node of the plasmon field pattern, where the near-field intensity is low. By adding 5 nm of Al_2O_3, the near-field antinode is shifted to the right junction, so more plasmons are split to end A, which agrees with the increased emission intensity of 633 nm wavelength at end A in the spectra in Fig. 6.32b. The wavelength dependence of the plasmon routing is determined by the plasmon field distributions for excitation light of different wavelengths. The routing behavior of SPPs of different wavelengths can be controlled separately by tuning the incident polarization without interfering with other wavelengths. Thus, the NW structure can be used for routing plasmons of multiple wavelengths simultaneously.

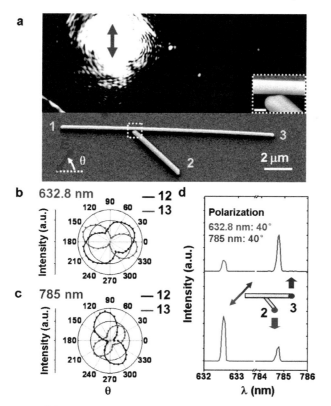

Figure 6.31 (a) Top: Optical image of a silver nanobranch excited by 633 nm wavelength laser light. The double-headed arrow represents the incident polarization. Bottom: SEM image of the nanobranch. The scale bar in the inset for the junction is 200 nm. θ is the rotation angle of the incident polarization. (b, c) Emission intensity from wire ends 2 and 3 as a function of the incident polarization angle for 633 and 785 nm wavelength excitation, respectively. (d) The spectra collected from wire ends 3 (upper curves) and 2 (lower curves). The polarization angle of the incident light is 40°. Reprinted with permission from Ref. [92]. Copyright (2010) American Chemical Society.

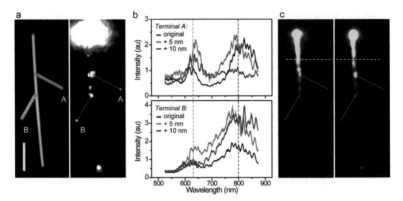

Figure 6.32 (a) A structure composed of three Ag NWs was illuminated by supercontinuum light with the incident polarization parallel to the main NW. The scale bar is 5 μm. (b) The upper panel shows the emission spectra at terminal A from the right branch for the original structure (black) and for 5 nm (red) and 10 nm (blue) of Al_2O_3 layer deposited. The lower panel is for terminal B from the left branch. (c) The QD fluorescence images when excited with a laser of 633 nm wavelength for the original structure (left) and after 5 nm more of Al_2O_3 was deposited (right). The dashed red lines in (a) and (c) show the positions of two branch wires. The dashed white line in (c) is a visual guide to show the shift of the near-field nodes and antinodes due to the addition of 5 nm of Al_2O_3. The batch of NWs used for this structure has a radius of about 150 nm. The original structure is coated by Al_2O_3 of 30 nm thickness [60].

The polarization dependence of the emission intensity at the terminals of nanobranches is sensitive to the thickness of the coated dielectric layer on the NW surface. For the branched structure in Fig. 6.33a, the emission intensities at terminals A and B show opposite dependence on the polarization of the input light, as shown in Fig. 6.33b. By depositing Al_2O_3 of 5 nm thickness, the polarization dependence of both A and B is changed (Fig. 6.33c). The polarization dependence of terminal B is reversed, that is, the polarization originally for the maximum output intensity produces the minimum output intensity. Depositing additional 5 nm of Al_2O_3 largely influences terminal A, resulting in reversed polarization dependence (comparing Figs. 6.33b and 6.33d). The scattering images before and after Al_2O_3 deposition in Fig. 6.33e clearly show the different outputs under the same excitation polarization, which is caused by the change of near-field distribution, as shown in Fig. 6.33f.

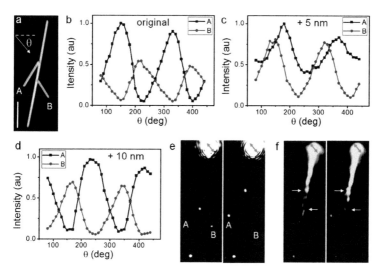

Figure 6.33 (a) White-light image of a NW network. The length of the scale bar is 5 μm. (b–d) Dependence of the output intensity at NW ends A and B on the polarization of the excitation light focused on the top end of the long wire for the original structure (b) and for 5 nm (c) and 10 nm (d) Al_2O_3 film deposited. The Al_2O_3 thickness of the original structure is 30 nm. The output intensities are normalized by the maximum values of the output intensity of terminal A. (e) Scattering images for the original structure and for 10 nm of Al_2O_3 deposited with the same incident polarization, as indicated by the double-headed arrows. (f) QD fluorescence images corresponding to (e). The white arrows in (f) are visual guides for the connection positions. Reproduced from Ref. [58] with permission from The Royal Society of Chemistry.

Besides incident polarization and dielectric coating, the conversion of plasmon modes induced by structural asymmetry can also influence the near-field distribution, as demonstrated in Fig. 6.17. So it can be used for manipulating the plasmon routing behavior as well [57]. For a branched silver NW structure with a silver nanoparticle, shown in Fig. 6.34a, simulation was performed by taking the TM_0 mode as input. By tuning the position of the nanoparticle (changing the distance d marked in Fig. 6.34a), the output intensities of the two branches are changed, as shown in Fig. 6.34b. For certain values of d, the power is routed to the top branch or the bottom branch, realizing the function of a single-pole double-throw switch. The electric field distributions in Fig. 6.34c,d show that the symmetric field distribution of the input TM_0 mode becomes zigzag shaped after passing the nanoparticle. By moving the nanoparticle along the NW, the maximum of the electric field intensity can be shifted from one side of the junction to the other side, which results in power switching between the two branches. The zigzag field distribution indicates the generation of the HE_1 mode in addition to the TM_0 mode. Here the presence of the nanoparticle breaks the structural symmetry of the NW, and scattering occurs at the nanoparticle. The scattering of localized modes in the nanogap and on the nanoparticle results in the symmetry change of the wave front, leading to the generation of the HE_1 mode. This mode conversion effect provides a way to manipulate plasmon modes at nanoscale and control plasmon propagation in nanocircuits.

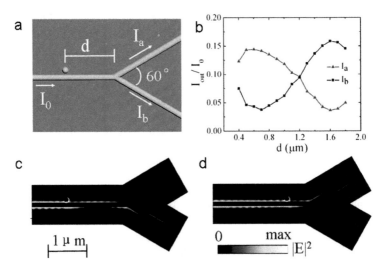

Figure 6.34 Nanoparticle-mediated mode conversion for a tunable switch in a branched Ag NW. (a) The sketch of the proposed switch structure. (b) The output intensities I_{out} at the two branches, normalized by the input intensity I_0 as a function of particle locations. (c, d) Electric field distributions for the two states of power switched between the two output branches. The radii of the NW and the particle are both 60 nm and their separation is 5 nm. The wavelength is 633 nm. The refractive index of the environmental medium is 1.56 [57].

As discussed in Section 6.3.6, due to the strong spin–orbit interaction occurring at the NW end, the incident light of opposite circular polarizations generates SPPs propagating along different trajectories on the NW. This effect can be used to sort light with different circular polarizations into separate physical paths. For a Y-shaped branched NW structure (Fig. 6.35a), the vertical NW is used as the main NW for exciting SPPs. Circularly polarized light focused at the end of the main NW will couple directionally to the propagating SPPs on the NW with periodic zigzag field patterns. By designing the length of the main NW as half the period, the field distribution at the junction will be asymmetric, and thus the SPPs will be directed to one of the two output branches. Figure 6.35b shows the simulated electric field intensity distributions for opposite circular polarizations. As can be seen, the SPPs excited by different polarizations propagate along different paths, realizing the spin sorting in the branched NW structure. The spin-sorting functionality is demonstrated in experiment (Fig. 6.35c). The fabricated Y-branch structure is immersed into index-matching oil to get a homogeneous dielectric environment. The polarization state of the excitation light is controlled by rotating a quarter-wave plate inserted into the path of the laser light. The SPPs excited by circularly polarized light with opposite spins (θ = 45° and 135° in Fig. 6.35d) are directed to different branches with the intensity ratio between two output ports higher than 6:1. These results demonstrate a novel approach to loading and sorting optical information, encoded in the spin degree of freedom of photons, in plasmonic nanocircuits.

The spectral splitting of SPPs in a silver NW can also be realized by making periodic corrugations on the NW, as shown in Fig. 6.36 [93]. For a silver NW of 170 nm diameter, two gratings with periods of 470 nm and 520 nm are fabricated by focused ion beam milling. The propagating SPPs will be modulated by the photonic band structure determined by the grating period. At the band gap ($n\pi/P$, where n is an integer and P is the period), the SPPs with specific frequencies are back-reflected, so they cannot propagate forward. By cutting single grooves on the NW (as shown in the squares marked by "v" in Fig. 6.36a), the SPPs can be coupled out as light. By focusing a polychromatic laser beam with multiple wavelengths onto the right end of the NW, SPPs are launched in the NW. Both the optical images and the spectra show that the frequencies of the light emitted at the three output ports (out-1 and out-2 correspond to the single grooves in Fig. 6.36a, and out-3 corresponds to the left end of the NW) are different. When the SPPs reach the first grating, the SPPs corresponding to 568 nm laser light are forbidden to propagate and scattered at the out-1 groove as yellow light. Grating 2 forbids the SPPs for wavelengths of 480–540 nm, which are scattered at the out-2 groove as green light. The remaining SPPs pass through grating 2 and couple out at the left end as red light.

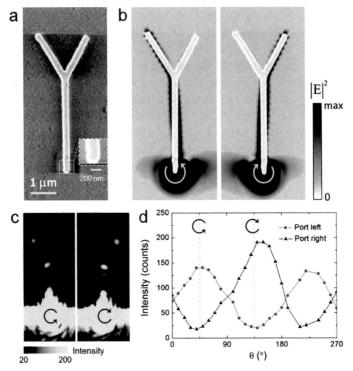

Figure 6.35 Spin-dependent directional propagation of SPPs in a gold branched NW. (a) SEM image of the fabricated gold branched NW structure. The gold NW is about 260 nm wide and 150 nm thick. The length of the main wire is about 3.7 μm. (b) The SEM image in (a) overlaid with the simulated near-field distributions for excitations of 785 nm wavelength with different circular polarizations denoted by the circular arrows. A uniform dielectric environment ($n = 1.5$) is used for the simulation, and the field distributions are extracted from the central horizontal section of the structure. (c) Experimental scattering images for excitation by focused laser beams of 785 nm wavelength with opposite circular polarizations denoted by the circular arrows. (d) Measured intensities from the two output ports of the branched NW structure for varying angle θ between the optical axis of the quarter-wave plate and the linear polarization of the laser beam. θ values of 0°, 90°, 180°, and 270° correspond to linear polarization parallel to the main NW of the Y-branch structure, while θ values of 45° and 135° correspond to circularly polarized light with opposite spins, as denoted by the circular arrows. Reprinted with permission from Ref. [76]. Copyright (2016) by the American Physical Society.

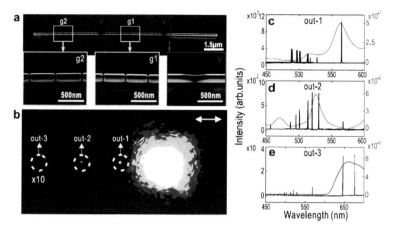

Figure 6.36 Selective propagation and emission of SPPs along a silver NW with cascading corrugation gratings. (a) SEM images of the silver NW with cascading gratings and the enlarged structure details: corrugations with $P = 520$ nm (left: grating 2), corrugations with $P = 470$ nm (middle: grating 1), and the "bus stop" groove (right: out-1). The NW diameter is about 170 nm. Both the width and the depth of the corrugation are 30 nm. (b) Emission micrograph of the structured NW illuminated by a polychromatic laser beam from the input end (rightmost end). The bright spots marked by the dashed white circles indicate the light scattered from grooves out-1, out-2 and out-3 (leftmost end). The white double-headed arrow indicates the polarization direction of the input beam. (c–e) Experimental (black lines) and calculated (curves) scattered light spectra from out-1, out-2, and out-3, respectively. Reprinted by permission from Macmillan Publishers Ltd: [*Scientific Reports*] (Ref. [93]), copyright (2013).

6.4.2 SPP Modulation, Logic Gates, and Computing

For a branched NW structure, besides the main wire end, the branch can be used as a second input end. Two plasmon waves can be launched independently from terminals I1 and I2 in the system shown in Fig. 6.37a. The polarization of the two laser beams is controlled independently, and the phase difference between the two laser beams can be tuned by a Babinet-Soleil compensator. When the phase difference between input I1 and I2 is increased monotonically, the output intensity at terminal O is varied in an oscillating manner, as shown in Fig. 6.37b. The strength of the plasmon interference can be described by visibility, defined as $(I_{max} - I_{min})/(I_{max} + I_{min})$, where I_{max} and I_{min} are the maximum and minimum values, respectively, of the output intensity at terminal O. The visibility is dependent on the polarization of the input light at terminal I1 [94]. Under optimal polarization, the visibility can be close to 1. Figure 6.37c shows the scattering images and corresponding QD fluorescence images for two phase differences. As can be seen, the interference influences both the near-field intensity and the output intensity at the NW end.

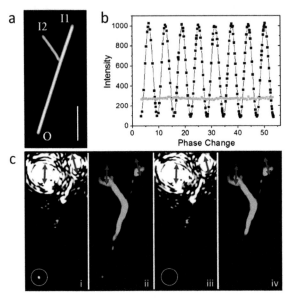

Figure 6.37 Interference of plasmons in Ag NWs. (a) Optical image of a simple two-NW network composed of a primary NW and a converging secondary input NW. The scale bar is 5 μm. (b) Scattering intensity at output terminal O as a function of the optical phase delay between I1 and I2 inputs . The horizontal line is the intensity for either I1 or I2 input. (c-i, c-iii) Scattering images for the case of two inputs I1 and I2, but with different phases. (c-ii, c-iv) QD fluorescence images corresponding to (c-i) and (c-iii), respectively. The double-headed arrows indicate the polarization of the input excitation laser of 633 nm wavelength. Reprinted with permission from Ref. [48]. Copyright (2011) American Chemical Society.

The difference of the maximum and minimum output intensity is so big that they can be assigned as ON and OFF states, or 1 and 0 states, respectively. For the simple NW network shown in Fig. 6.37, either I1 or I2 input results in a 1 output, while both I1 and I2 inputs result in a 1 output for constructive interference. Thus this three-terminal NW structure works as an OR gate. By setting a threshold intensity larger than the output intensity with only one input, this structure can work as an AND gate for constructive interference. For example, if the threshold intensity is set as 600, the output state is 0 for either I1 or I2 input and the output state is 1 for both I1 and I2 inputs, which corresponds to the AND operation. For destructive interference, either I1 or I2 input results in a 1 output, while both I1 and I2 inputs result in a 0 output. This corresponds to an XOR gate. If one of the inputs is used as the control signal, the input 0 is inverted to 1 and 1 is inverted to 0, so the structure realizes the function of the NOT gate.

For a more complex NW network with two input ends and two output ends, shown in Fig. 6.38a, the plasmon interference with varied phase difference can determine the plasmon routing to the two output ends. Figure 6.38b shows the changes of the output intensity in one interference cycle. The plasmon energy can be routed to the O1 terminal for certain input phases (Fig. 6.38b-i). When the phase difference is increased, part of the energy is split to the O2 terminal (Fig. 6.38b-ii). Further increase of the phase difference will make the plasmons routed to the O2 terminal (Fig. 6.38b-iii). When the phase is increased further, plasmons are split to the two output paths (Fig. 6.38b-iv) and finally switched totally to the O1 terminal (Fig. 6.38b-v), the same as the beginning of the interference cycle.

Figure 6.38c shows the QD fluorescence images corresponding to Fig. 6.38b. The near-field distribution revealed by QD fluorescence explains the output behavior in Fig. 6.38b. The general intensity of the electric field on the two arms of the outputs is switched between the strong and the weak in an alternating way, which determines the output intensity switching at the two output terminals O1 and O2. The dashed rectangle in Fig. 6.38c highlights the junction between the output branch wire and the main wire. As can be seen, the intensity at the junction is strongly related to the plasmon transmission in the NW network. When the intensity at the junction is weak, the plasmons propagate along the main wire (Fig. 6.38c-i). When the intensity at the junction is strong, most energy is switched to the branch wire (Fig. 6.38c-iii). Here the control on the local field intensity at the junction is controlled by the relative phase of the two input light beams. As we have discussed earlier, the incident polarization plays a critical role in determining the plasmon routing behavior in joint NW structures. The phase is another parameter strongly influencing the near-field distribution in the NW networks. The combination of polarization and phase can make more output behaviors that are not achievable by tuning only the polarization or phase.

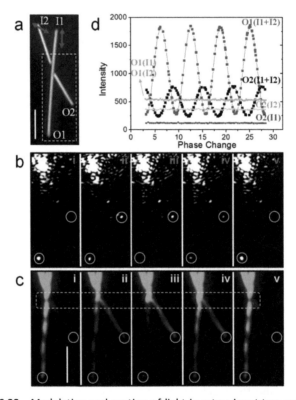

Figure 6.38 Modulation and routing of light in a two-input-two-output NW network. (a) Optical image of the network. (b) Scattering images for two-beam interference in one cycle. (c) QD fluorescence images in one interference cycle. (d) Scattering intensities at O1 and O2 terminals. Red: intensity of O1 for simultaneous input of both I1 and I2; black; intensity of O2 for simultaneous input of both I1 and I2; green: intensity of O1 for I1 only and I2 only; cyan: intensity of O2 for I2 only; blue: intensity of O2 for I1 only. Al_2O_3 thickness is 50 nm. The scale bar is 5 μm. Double-headed arrows in (a) show the polarizations of the two laser beams of 633 nm wavelength, and the dashed white rectangle in (a) marks the area displayed in (b) and (c). The yellow dashed rectangle in (c) highlights the variation of the electric field intensity at the junction. Reprinted with permission from Ref. [48]. Copyright (2011) American Chemical Society.

Figure 6.38d shows the output intensities at O1 and O2 terminals for different inputs in a few interference cycles. By defining a threshold intensity of 450 for 1 and 0 states, this simple four-terminal NW network can function as a binary half adder. Two inputs of I1 and I2 result in 1 output at O2 and 0 output at O1: 1 + 1 = 10, while one input of only I1 or only I2 results in 0 at O2 and 1 at O1: 1 + 0 = 0 + 1 = 01. It is noted that the half adder is realized due to the multiple plasmon modes in the NW, which enable destructive interference at O1 and constructive interference at O2. A few examples of NW-based logic gates are listed in Table 6.1 [48]. In simple NW networks, a complete set of Boolean logic gates can be realized. These results show the potential of using plasmonic waveguides for information processing in nanophotonic circuits.

Table 6.1 Examples of all-optical logic operations based on NW networks[a]

AND	0,0 empty 0	0,1 0	1,0 0	1,1 1
OR	0,0 = 0	0,1 = 1	1,0 = 1	1,1 = 1
XOR	0,0 = 0	0,1 = 1	1,0 = 1	1,1 = 0
NOT	1 control, 0 = 1	1, 1 = 0		
NAND	0,0 empty 1 control 1	0,1 1	1,0 1	1,1 0
Half Adder	0,0 → 0, 0	0,1 → 1, 0	1,0 → 0, 1	1,1 → 0, 1
	0+0=(0 0)	0+1=(0 1)	1+0=(0 1)	1+1=(1 0)

[a]The numbers on the left are inputs, and the numbers on the right are outputs. Unused terminals are labeled "empty." The terminals labeled "control" require the input to be ON. The structure for AND can be simplified as the structure for OR.
Source: Reprinted with permission from Ref. [48]. Copyright (2011) American Chemical Society.

To realize more complex logic functions, the elementary logic gates need to be cascaded. The cascade of plasmon-based interferometric logic gates was investigated by examining one of the universal logic gates, the NOR gate, as shown in Fig. 6.39. A structure composed of three NWs is designed for the cascaded NOR gate. The left part of the main wire (longest) and the left branch wire combine to function as the OR gate, while the right part of the main wire and the right branch wire are for the NOT gate. The ends labeled I1 and I2 are the two input terminals, and the end labeled C is for the control signal, that is, one of the inputs of the NOT gate.

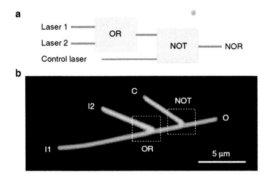

Figure 6.39 (a) Schematic illustration of the logic gate NOR built by cascaded OR and NOT gates. (b) Optical image of the designed Ag NW structure. Reprinted by permission from Macmillan Publishers Ltd: [*Nature Communications*] (Ref. [95]), copyright (2011).

Table 6.2 Outputs of logic gates [95]

Input			Output		
I1	I2	C	O1 = I1 OR I2	(O1, C) = NOT O1	O = I1 NOR I2
0	0	1	0	1	1
0	1	1	1	0	0
1	0	1	1	0	0
1	1	1	1	0	0

In experiments, the laser light was split into three beams focused onto the three input terminals. The scattering images in Fig. 6.40 show that in the absence of the control signal, the structure operates as an OR gate (Fig. 6.40a). When the control signal is enabled (Fig. 6.40b), the NOT operation inverts the results of the OR operation. Figure 6.40c shows the results corresponding to the NOT function acting on the input conditions in the middle and bottom panels in

Fig. 6.40a. As can be seen, the NOT operation successfully inverted the OR operation results. Therefore, the cascade was realized. Table 6.2 summarizes the outputs of the logic gates.

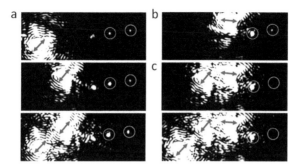

Figure 6.40 (a) The scattering images showing OR operations. (b) The scattering image when only the control signal C was enabled. (c) The scattering images for NOR operations corresponding to a NOT function acting on the middle and bottom panels in (a). The double-headed arrows show the incident laser polarization. The circles mark the C-branch junction and the O terminal. Reprinted by permission from Macmillan Publishers Ltd: [*Nature Communications*] (Ref. [95]), copyright (2011).

To explore the mechanism of the plasmon-interference-based cascaded logic gates, the QD fluorescence imaging technique was employed to detect the near-field distribution in the NOR gate device. Figures 6.41b–d show the QD fluorescence images for three different input conditions. The common characteristic of these three images is that the local electric field intensity at the junction of the control branch is locally maximal. This strong local field guarantees that the control signal can interfere strongly with the signal from the OR operation. Since the local electric field intensity distribution is dependent on the incident polarization, the device works when the polarization of the input signal makes the electric field intensity at the control branch junction locally strong. If the polarization in Fig. 6.41b is rotated 90°, the antinode in the near-field distribution pattern at the control branch junction will be shifted to the opposite side of the junction. Hence the control signal cannot interfere strongly with the I1 input and the NOR operation cannot be realized.

The results obtained in the studies of plasmon routing, switching, and interference-based signal processing show that the near-field distribution resulting from the multiple SPP modes plays a central role in determining the behavior of the NW-based devices. The manipulation to the near field is essential for designing various device functions. The control on the near-field distribution provides a viable way to control the interaction between devices in plasmonic circuits and can serve as a design rule to construct more complex systems. Combining with special materials with controllable dielectric properties, it's possible to actively control the signal transmitting and processing in the nanocircuits.

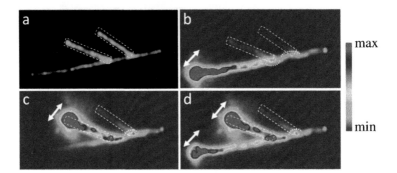

Figure 6.41 (a) QD fluorescence image by wide-field illumination. Dashed white lines show the outline of the branch NWs. (b–d) QD fluorescence images with excitation at different terminals, with polarization indicated by the white arrows. Reprinted by permission from Macmillan Publishers Ltd: [*Nature Communications*] (Ref. [95]), copyright (2011).

6.4.3 Hybrid Plasmonic-Photonic Nanowire Devices

Plasmonic NW waveguides can support light propagation with tight field confinement; meanwhile they suffer high energy losses. Thus, it's challenging to build on-chip integrated nanophotonic circuits with only metal nanowaveguides. The trade-off between mode size and propagation loss must be considered carefully. The hybridization with photonic (dielectric or semiconductor) NWs is a possible strategy to decrease the energy loss during propagation, since the photonic waveguides have much lower loss in the optical frequency. Guo et al. experimentally investigated the coupling between Ag NWs and ZnO NWs to build hybrid nanophotonic components [31]. As shown in Fig. 6.42a, the excitation of plasmons is realized by using a tapered fiber. The light of 650 nm wavelength from the nanofiber is coupled to the Ag NW, which has an overlap of less than 1 µm with the nanofiber. The emission from the right terminal of the Ag NW indicates the generation of propagating SPPs (Fig. 6.42b). For connected ZnO and Ag NWs (Fig. 6.42d), the light from the nanofiber transmits through the ZnO NW and then excites the propagating plasmons in the Ag NW (Fig. 6.42c). On the contrary, the SPPs launched in the Ag NW by the nanofiber can also couple to the ZnO NW (Fig. 6.42e–g). In a branched structure composed of a ZnO NW and a Ag NW, the coupling efficiency from the ZnO NW into the Ag NW at 650 nm wavelength was demonstrated to be about 82% after the propagation loss of the Ag NW was calibrated. By using coupled Ag NWs and ZnO NWs, a Mach–Zehnder interferometer and a microring cavity were demonstrated as well.

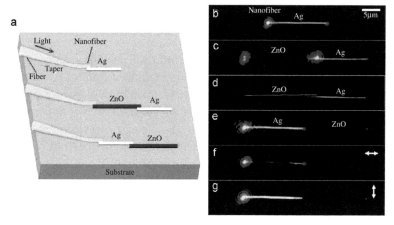

Figure 6.42 Direct coupling of plasmonic and photonic NWs. (a) Schematic illustration of light coupling between silica nanofibers, ZnO NWs, and Ag NWs. Light from the nanofiber can excite plasmons in the Ag NW directly (top) or via an intervening ZnO NW (middle). Plasmons in the Ag NW can convert back to light in the ZnO NW (bottom). (b, c, e) Experimental observations corresponding to the three different coupling schemes illustrated in (a). (d) SEM image of the coupler shown in (c). The diameter of the ZnO NW is 340 nm, and the diameter of the Ag NW is 320 nm. (f, g) Optical images taken with the polarization along and perpendicular to the NWs shown in (e), respectively. Reprinted with permission from Ref. [31]. Copyright (2009) American Chemical Society.

A theoretical study shows that by using a butt-coupling scheme, the conversion efficiency of plasmons in a metal NW to photons in a dielectric fiber can be as high as 95% in the visible light range and close to 100% in the near-infrared range [96]. Coupling of light into Ag NWs from polymer waveguides and SnO_2 ribbons was experimentally studied [97, 98]. By coupling a Ag NW with a high-gain CdSe NW, a hybrid photon-plasmon NW laser was realized [99]. In the hybrid waveguide of a Ag NW coupled to a fiber taper, the transmission of photonic quantum polarization entanglement was demonstrated [100]. The integration of plasmonic and photonic waveguides provides a way to overcome the limitation of high loss for the plasmonic waveguides and may promote the development of on-chip integrated nanophotonic circuits.

6.5 Summary

Metal NWs supporting propagating SPPs are the fundamental building block for nanophotonic integrated circuits. The SPPs on the NWs can be excited and detected by optical or electrical means. The multiple SPP modes on the NWs determine the peculiar propagation and emission behaviors. The excitation of multiple modes results in the periodic distribution of the electric field on the NWs, which plays a critical role in determining the SPP propagation in NW networks. By tuning the incident polarization, the dielectric environment, or the structural symmetry, the field distribution is changed, which can be conveniently used to manipulate the SPPs in the nanocircuits. The SPPs maintain the coherence of the excitation light, which can also be used to modulate the SPP propagation and emission. On the basis of the controls to SPPs, nanophotonic devices can be realized in NW structures. The SPPs can interact with dielectric and active optical materials, which can be used to realize various functions and improve the performances of plasmonic devices. Plasmonic NW waveguides and circuits provide a versatile platform for studying the light manipulation and light–matter interaction at nanoscale dimensions and may find applications in optical communication, information, quantum photonics, nanospectroscopy, and sensing.

References

1. E. Ozbay, *Science*, **311**, 189–193 (2006).
2. H. Wei, and H. X. Xu, *Nanophotonics*, **1**, 155–169 (2012).
3. L. Wendler, and R. Haupt, *J. Appl. Phys.*, **59**, 3289–3291 (1986).

4. J. Takahara, S. Yamagishi, H. Taki, A. Morimoto, and T. Kobayashi, *Opt. Lett.*, **22**, 475–477 (1997).
5. M. Quinten, A. Leitner, J. R. Krenn, and F. R. Aussenegg, *Opt. Lett.*, **23**, 1331–1333 (1998).
6. S. A. Maier, P. G. Kik, H. A. Atwater, S. Meltzer, E. Harel, B. E. Koel, and A. A. G. Requicha, *Nat. Mater.*, **2**, 229–232 (2003).
7. Z. P. Li, and H. X. Xu, *J. Quant. Spectrosc. Radiat. Transfer*, **103**, 394–401 (2007).
8. P. Berini, *Opt. Lett.*, **24**, 1011–1013 (1999).
9. P. Berini, *Phys. Rev. B*, **61**, 10484–10503 (2000).
10. J. C. Weeber, A. Dereux, C. Girard, J. R. Krenn, and J. P. Goudonnet, *Phys. Rev. B*, **60**, 9061–9068 (1999).
11. R. Zia, M. D. Selker, and M. L. Brongersma, *Phys. Rev. B*, **71**, 165431 (2005).
12. J. C. Weeber, J. R. Krenn, A. Dereux, B. Lamprecht, Y. Lacroute, and J. P. Goudonnet, *Phys. Rev. B*, **64**, 045411 (2001).
13. J. C. Weeber, M. U. Gonzalez, A. L. Baudrion, and A. Dereux, *Appl. Phys. Lett.*, **87**, 221101 (2005).
14. R. Zia, J. A. Schuller, and M. L. Brongersma, *Phys. Rev. B*, **74**, 165415 (2006).
15. I. V. Novikov, and A. A. Maradudin, *Phys. Rev. B*, **66**, 035403 (2002).
16. D. F. P. Pile, and D. K. Gramotnev, *Opt. Lett.*, **29**, 1069–1071 (2004).
17. S. I. Bozhevolnyi, V. S. Volkov, E. Devaux, J. Y. Laluet, and T. W. Ebbesen, *Nature,* **440**, 508–511 (2006).
18. D. F. P. Pile, T. Ogawa, D. K. Gramotnev, T. Okamoto, M. Haraguchi, M. Fukui, and S. Matsuo, *Appl. Phys. Lett.*, **87**, 061106 (2005).
19. D. F. P. Pile, T. Ogawa, D. K. Gramotnev, Y. Matsuzaki, K. C. Vernon, K. Yamaguchi, T. Okamoto, M. Haraguchi, and M. Fukui, *Appl. Phys. Lett.*, **87**, 261114 (2005).
20. K. Tanaka, and M. Tanaka, *Appl. Phys. Lett.*, **82**, 1158–1160 (2003).
21. B. Steinberger, A. Hohenau, H. Ditlbacher, A. L. Stepanov, A. Drezet, F. R. Aussenegg, A. Leitner, and J. R. Krenn, *Appl. Phys. Lett.*, **88**, 094104 (2006).
22. M. Z. Alam, J. Meier, J. S. Aitchison, and M. Mojahedi, *Conference on Lasers & Electro-Optics/Quantum Electronics and Laser Science Conference*, 2638–2639 (2007).
23. R. F. Oulton, V. J. Sorger, D. A. Genov, D. F. P. Pile, and X. Zhang, *Nat. Photonics,* **2**, 496–500 (2008).

24. H. Raether, *Surface Plasmons on Smooth and Rough Surfaces and on Gratings* (Springer, 1988).
25. M. Allione, V. V. Temnov, Y. Fedutik, U. Woggon, and M. V. Artemyev, *Nano Lett.*, **8**, 31–35 (2008).
26. A. W. Sanders, D. A. Routenberg, B. J. Wiley, Y. N. Xia, E. R. Dufresne, and M. A. Reed, *Nano Lett.*, **6**, 1822–1826 (2006).
27. M. W. Knight, N. K. Grady, R. Bardhan, F. Hao, P. Nordlander, and N. J. Halas, *Nano Lett.*, **7**, 2346–2350 (2007).
28. S. P. Zhang, C. Z. Gu, and H. X. Xu, *Small*, **10**, 4264–4269 (2014).
29. Z. Y. Fang, L. R. Fan, C. F. Lin, D. Zhang, A. J. Meixner, and X. Zhu, *Nano Lett.*, **11**, 1676–1680 (2011).
30. B. Hecht, H. Bielefeldt, L. Novotny, Y. Inouye, and D. W. Pohl, *Phys. Rev. Lett.*, **77**, 1889–1892 (1996).
31. X. Guo, M. Qiu, J. M. Bao, B. J. Wiley, Q. Yang, X. N. Zhang, Y. G. Ma, H. K. Yu, and L. M. Tong, *Nano Lett.*, **9**, 4515–4519 (2009).
32. W. H. Wang, Q. Yang, F. R. Fan, H. X. Xu, and Z. L. Wang, *Nano Lett.*, **11**, 1603–1608 (2011).
33. N. Liu, Z. P. Li, and H. X. Xu, *Small*, **8**, 2641–2646 (2012).
34. H. Ditlbacher, A. Hohenau, D. Wagner, U. Kreibig, M. Rogers, F. Hofer, F. R. Aussenegg, and J. R. Krenn, *Phys. Rev. Lett.*, **95**, 257403 (2005).
35. Q. Li, H. Wei, and H. X. Xu, *Chin. Phys. B*, **23**, 097302 (2014).
36. P. Bharadwaj, A. Bouhelier, and L. Novotny, *Phys. Rev. Lett.*, **106**, 226802 (2011).
37. D. M. Koller, A. Hohenau, H. Ditlbacher, N. Galler, F. Reil, F. R. Aussenegg, A. Leitner, E. J. W. List, and J. R. Krenn, *Nat. Photonics*, **2**, 684–687 (2008).
38. P. Neutens, L. Lagae, G. Borghs, and P. Van Dorpe, *Nano Lett.*, **10**, 1429–1432 (2010).
39. K. C. Y. Huang, M. K. Seo, T. Sarmiento, Y. J. Huo, J. S. Harris, and M. L. Brongersma, *Nat. Photonics*, **8**, 244–249 (2014).
40. A. Babuty, A. Bousseksou, J. P. Tetienne, I. M. Doyen, C. Sirtori, G. Beaudoin, I. Sagnes, Y. De Wilde, and R. Colombelli, *Phys. Rev. Lett.*, **104**, 226806 (2010).
41. R. J. Walters, R. V. A. van Loon, I. Brunets, J. Schmitz, and A. Polman, *Nat. Mater.*, **9**, 21–25 (2010).
42. P. Y. Fan, C. Colombo, K. C. Y. Huang, P. Krogstrup, J. Nygard, A. F. I. Morral, and M. L. Brongersma, *Nano Lett.*, **12**, 4943–4947 (2012).

43. J. Li, H. Wei, H. Shen, Z. X. Wang, Z. S. Zhao, X. M. Duan, and H. X. Xu, *Nanoscale*, **5**, 8494–8499 (2013).
44. P. Rai, N. Hartmann, J. Berthelot, J. Arocas, G. C. des Francs, A. Hartschuh, and A. Bouhelier, *Phys. Rev. Lett.*, **111**, 026804 (2013).
45. W. Cai, R. Sainidou, J. J. Xu, A. Polman, and F. J. G. de Abajo, *Nano Lett.* **9**, 1176–1181 (2009).
46. D. Rossouw, and G. A. Botton, *Phys. Rev. Lett.*, **110**, 066801 (2013).
47. A. Drezet, A. Hohenau, J. R. Krenn, M. Brun, and S. Huant, *Micron*, **38**, 427–437 (2007).
48. H. Wei, Z. P. Li, X. R. Tian, Z. X. Wang, F. Z. Cong, N. Liu, S. P. Zhang, P. Nordlander, N. J. Halas, and H. X. Xu, *Nano Lett.*, **11**, 471–475 (2011).
49. H. Wei, D. Ratchford, X. Q. Li, H. X. Xu, and C. K. Shih, *Nano Lett.*, **9**, 4168–4171 (2009).
50. A. L. Falk, F. H. L. Koppens, C. L. Yu, K. Kang, N. D. Snapp, A. V. Akimov, M. H. Jo, M. D. Lukin, and H. Park, *Nat. Phys.*, **5**, 475–479 (2009).
51. K. M. Goodfellow, C. Chakraborty, R. Beams, L. Novotny, and A. N. Vamiyakas, *Nano Lett.*, **15**, 5477–5481 (2015).
52. P. Neutens, P. Van Dorpe, I. De Vlaminck, L. Lagae, and G. Borghs, *Nat. Photonics,* **3**, 283–286 (2009).
53. N. Ittah, and Y. Selzer, *Nano Lett.*, **11**, 529–534 (2011).
54. D. E. Chang, A. S. Sorensen, P. R. Hemmer, and M. D. Lukin, *Phys. Rev. B*, **76**, 035420 (2007).
55. S. P. Zhang, H. Wei, K. Bao, U. Hakanson, N. J. Halas, P. Nordlander, and H. X. Xu, *Phys. Rev. Lett.*, **107**, 096801 (2011).
56. D. Pan, H. Wei, and H. X. Xu, *Chin. Phys. B*, **22**, 097305 (2013).
57. D. Pan, H. Wei, Z. L. Jia, and H. X. Xu, *Sci. Rep.*, **4**, 4993 (2014).
58. H. Wei, D. Pan, and H. X. Xu, *Nanoscale*, **7**, 19053–19059 (2015).
59. L. M. Tong, H. Wei, S. P. Zhang, and H. X. Xu, *Sensors*, **14**, 7959–7973 (2014).
60. H. Wei, S. P. Zhang, X. R. Tian, and H. X. Xu, *Proc. Natl. Acad. Sci. USA*, **110**, 4494–4499 (2013).
61. Z. X. Wang, H. Wei, D. Pan, and H. X. Xu, *Laser Photonics Rev.*, **8**, 596–601 (2014).
62. V. V. Temnov, U. Woggon, J. Dintinger, E. Devaux, and T. W. Ebbesen, *Opt. Lett.*, **32**, 1235–1237 (2007).
63. C. Rewitz, T. Keitzl, P. Tuchscherer, J. Huang, P. Geisler, G. Razinskas, B. Hecht, and T. Brixner, *Nano Lett.*, **12**, 45–49 (2012).

64. Y. G. Ma, X. Y. Li, H. K. Yu, L. M. Tong, Y. Gu, and Q. H. Gong, *Opt. Lett.*, **35**, 1160–1162 (2010).
65. T. Shegai, Y. Z. Huang, H. X. Xu, and M. Käll, *Appl. Phys. Lett.*, **96**, 103114 (2010).
66. Z. P. Li, K. Bao, Y. R. Fang, Z. Q. Guan, N. J. Halas, P. Nordlander, and H. X. Xu, *Phys. Rev. B*, **82**, 241402 (2010).
67. S. P. Zhang, and H. X. Xu, *ACS Nano*, **6**, 8128–8135 (2012).
68. Y. S. Bian, and Q. H. Gong, *Nanoscale*, **7**, 4415–4422 (2015).
69. Z. P. Li, F. Hao, Y. Z. Huang, Y. R. Fang, P. Nordlander, and H. X. Xu, *Nano Lett.*, **9**, 4383–4386 (2009).
70. M. A. Lieb, J. M. Zavislan, and L. Novotny, *J. Opt. Soc. Am. B*, **21**, 1210–1215 (2004).
71. T. Shegai, V. D. Miljkovic, K. Bao, H. X. Xu, P. Nordlander, P. Johansson, and M. Käll, *Nano Lett.*, **11**, 706–711 (2011).
72. Z. L. Jia, H. Wei, D. Pan, and H. X. Xu, *Nanoscale*, **8**, 20118–20124 (2016).
73. H. Wei, X. R. Tian, D. Pan, L. Chen, Z. L. Jia, and H. X. Xu, *Nano Lett.*, **15**, 560–564 (2015).
74. Z. P. Li, K. Bao, Y. R. Fang, Y. Z. Huang, P. Nordlander, and H. X. Xu, *Nano Lett.*, **10**, 1831–1835 (2010).
75. R. Mignani, E. Recami, and M. Baldo, *Lettere Al Nuovo Cimento*, **11**, 568–572 (1974).
76. D. Pan, H. Wei, L. Gao, and H. X. Xu, *Phys. Rev. Lett.*, **117**, 166803 (2016).
77. H. Wei, F. Hao, Y. Z. Huang, W. Z. Wang, P. Nordlander, and H. X. Xu, *Nano Lett.*, **8**, 2497–2502 (2008).
78. Y. R. Fang, H. Wei, F. Hao, P. Nordlander, and H. X. Xu, *Nano Lett.*, **9**, 2049–2053 (2009).
79. D. E. Chang, A. S. Sorensen, P. R. Hemmer, and M. D. Lukin, *Phys. Rev. Lett.*, **97**, 053002 (2006).
80. D. E. Chang, A. S. Sorensen, E. A. Demler, and M. D. Lukin, *Nat. Phys.*, **3**, 807–812 (2007).
81. A. V. Akimov, A. Mukherjee, C. L. Yu, D. E. Chang, A. S. Zibrov, P. R. Hemmer, H. Park, and M. D. Lukin, *Nature*, **450**, 402–406 (2007).
82. R. Kolesov, B. Grotz, G. Balasubramanian, R. J. Stohr, A. A. L. Nicolet, P. R. Hemmer, F. Jelezko, and J. Wrachtrup, *Nat. Phys.*, **5**, 470–474 (2009).
83. S. D. Liu, M. T. Cheng, Z. J. Yang, and Q. Q. Wang, *Opt. Lett.*, **33**, 851–853 (2008).

84. A. Huck, S. Kumar, A. Shakoor, and U. L. Anderson, *Phys. Rev. Lett.*, **106**, 096801 (2011).
85. S. Kumar, A. Huck, and U. L. Andersen, *Nano Lett.*, **13**, 1221–1225 (2013).
86. S. Kumar, N. I. Kristiansen, A. Huck, and U. L. Andersen, *Nano Lett.*, **14**, 663–669 (2014).
87. A. Gonzalez-Tudela, D. Martin-Cano, E. Moreno, L. Martin-Moreno, C. Tejedor, and F. J. Garcia-Vidal, *Phys. Rev. Lett.*, **106**, 020501 (2011).
88. G. Y. Chen, N. Lambert, C. H. Chou, Y. N. Chen, and F. Nori, *Phys. Rev. B*, **84**, 045310 (2011).
89. J. Yang, G. W. Lin, Y. P. Niu, and S. Q. Gong, *Opt. Express*, **21**, 15618–15626 (2013).
90. Q. Li, H. Wei, and H. X. Xu, *Nano Lett.*, **14**, 3358–3363 (2014).
91. Q. Li, H. Wei, and H. X. Xu, *Nano Lett.*, **15**, 8181–8187 (2015).
92. Y. R. Fang, Z. P. Li, Y. Z. Huang, S. P. Zhang, P. Nordlander, N. J. Halas, and H. X. Xu, *Nano Lett.*, **10**, 1950–1954 (2010).
93. Q. Hu, D. H. Xu, Y. Zhou, R. W. Peng, R. H. Fan, N. X. Fang, Q. J. Wang, X. R. Huang, and M. Wang, *Sci. Rep.*, **3**, 3095 (2013).
94. H. Wei, and H. X. Xu, *Nanoscale*, **4**, 7149–7154 (2012).
95. H. Wei, Z. X. Wang, X. R. Tian, M. Käll, and H. X. Xu, *Nat. Commun.*, **2**, 387 (2011).
96. X. W. Chen, V. Sandoghdar, and M. Agio, *Nano Lett.*, **9**, 3756–3761 (2009).
97. A. L. Pyayt, B. Wiley, Y. N. Xia, A. Chen, and L. Dalton, *Nat. Nanotechnol.*, **3**, 660–665 (2008).
98. R. X. Yan, P. Pausauskie, J. X. Huang, and P. D. Yang, *Proc. Natl. Acad. Sci. USA*, **106**, 21045–21050 (2009).
99. X. Q. Wu, Y. Xiao, C. Meng, X. N. Zhang, S. L. Yu, Y. P. Wang, C. X. Yang, X. Guo, C. Z. Ning, and L. M. Tong, *Nano Lett.*, **13**, 5654–5659 (2013).
100. M. Li, C. L. Zou, X. F. Ren, X. Xiong, Y. J. Cai, G. P. Guo, L. M. Tong, and G. C. Guo, *Nano Lett.*, **15**, 2380–2384 (2015).

Chapter 7

Gain-Assisted Surface Plasmon Resonances and Propagation

Ning Liu[a] and Hongxing Xu[b]

[a]*Department of Physics and Bernal Institute, University of Limerick, Limerick, Republic of Ireland*
[b]*School of Physics and Technology, and Institute for Advanced Studies, Wuhan University, Wuhan 430072, China*
ning.liu@ul.ie; hxxu@whu.edu.cn

7.1 Introduction

Surface plasmons (SPs) have attracted great attention in recent years due to their potential applications in various fields of science and technology [1, 2]. The capability to confine light into a subwavelength spatial dimension through plasmonic structures makes manipulating light at nanoscale possible. Various novel phenomena such as surface-enhanced fluorescence and Raman scattering, SP resonance–induced optical force enhancement, and extraordinary transmission through metallic nanoarrays have been discovered due to the excitation of SPs. More details of these fascinating properties of SPs can be found in other chapters of this

---

Nanophotonics: Manipulating Light with Plasmons
Edited by Hongxing Xu
Copyright © 2018 Pan Stanford Publishing Pte. Ltd.
ISBN 978-981-4774-14-7 (Hardcover), 978-1-315-19661-9 (eBook)
www.panstanford.com

book. In this chapter, we will focus on the behavior of plasmonic structures in the close vicinity of gain media.

Because of their ability to confine light into a subwavelength spatial dimension, as well as their fast propagation speed (up to 100 THz), nanoplasmonic structures have been regarded as one of the most promising candidates for next-generation information communication technology. They can either be used as building blocks to realize all-optical integrated photonic chip or be integrated with the silicon electronics in a fully compatible manner to solve some critical issues in current computer architectures such as high-speed interconnect at both intra- and interchip levels. With the surge of research on plasmonic circuits in recent years, both passive and active plasmonic components have been demonstrated, such as plasmonic waveguides, wavelength-dependent plasmonic routers [3], plasmon-based modulators, single-plasmon sources [4], and even plasmonic nanowire networks to perform logic algorithms [5, 6].

However, metallic loss that accompanies all plasmonic devices has hindered further integration of plasmonic-based components into practical circuits. In the most recent years, researchers have tried to partially replace or completely replace the normal dielectric part of the plasmonic devices with gain materials in order to compensate for the Ohmic loss caused by the metallic components within the devices [7–9]. Great successes have been achieved in the areas of amplification of long-range surface plasmon polaritons (LRSPPs) using fluorescent materials [10, 11], stimulated emission from metal-dielectric core-shell nanoparticles [12], and stimulated emission from hybrid SP modes at the metal–insulator–semiconductor interfaces [13–15]. Our group has demonstrated that hybrid plasmonic signals of different polarizations can be amplified on the same device through the pump-probe ultrafast technique [16].

The stimulated emission from nanoplasmonic structures is the evidence for the full compensation on the metallic loss in the nanoplasmonic structures by gain media. In the case of localized SP resonances, the critical gain coefficient—lasing threshold—is estimated to be in the range of 1000 cm^{-1}, a number much larger than that of small fluorescent molecules. This number is at the upper limit of what dye molecules or conjugated polymer materials can offer and is more achievable with semiconductor materials. The realization of signal amplification in long-range SPPs is given in Section 7.2. The results for lasing behavior obtained from the enhanced localized SP resonance is discussed in details in Section 7.3. For nonresonant propagating SPPs, the compensation for metallic Ohmic loss can be expressed as stimulated emission at specific cavity modes or the amplification of input signal. The stimulated emission from hybrid SP modes at the metal–insulator–semiconductor interfaces shows a critical gain exceeding 10,000 cm^{-1} and will be detailed in Section 7.4. At the end of this section, we will explain our approach to the realization of hybrid plasmonic signal amplification with the pump-probe technique. We will then finish this chapter with an outlook on the future possibilities of gain-assisted plasmonic devices.

7.2 Amplification of Long-Range Surface Plasmon Polaritons

To minimize the metallic loss that is associated with the SPP modes, some researchers sacrifice the strong mode confinement offered by short-range SPPs and trade it for low propagation loss. The result of this compromise is a mode called LRSPP. To obtain LRSPPs, a sufficiently thin metal film needs to be buried in the dielectric. The mathematical description of the LRSPP can be relatively simple. For any SPP mode existing at the interface of two infinite half media, metal and dielectric, its propagation constant β (parallel to the interface) is written as $\beta = \frac{\omega}{c}\sqrt{\frac{\varepsilon_m \varepsilon_d}{\varepsilon_m + \varepsilon_d}}$, where ω and c are the frequency and speed of electromagnetic waves in vacuum, respectively, and ε_d and ε_m are the permittivities of the dielectric and metal, respectively. When a thin metal film is embedded in dielectric, the two identical SPP modes at the upper and lower interfaces will couple together and form symmetrical and antisymmetrical modes, whose propagation constants parallel to the interface are implicitly described in the dispersion relation [2]:

$$\tanh(S_m t) = -\frac{2\varepsilon_d S_d \varepsilon_m S_m}{\varepsilon_d^2 S_m^2 + \varepsilon_m^2 S_d^2}, \; S_d = \sqrt{\beta^2 - \varepsilon_d k_0^2}, \; S_m = \sqrt{\beta^2 - \varepsilon_m k_0^2}, \quad (7.1)$$

where t is the thickness of the metal film and $k_0 = \omega/c$ is the wave number of light in vacuum. For the symmetric mode, most of the mode profile is within the cladding dielectric layer and is only weakly bound to the metal film. With the decrease in the metal thickness, the part of the mode field within the metal also becomes progressively smaller, resulting in drastically decreased mode absorption and propagation loss. In general, the propagation length of LRSPPs can increase more than an order of magnitude compared to that of a short-range SPP mode.

Since the propagation loss for the LRSPP is usually under 100 cm^{-1}, which is feasible with the highly fluorescent dye molecules or conjugated polymers, the experimental efforts are mostly focused on the construction of symmetric (refractive-index-matched) active dielectric layers at the top and bottom interfaces of the metal stripe. Gather et al. adapted the dielectric-metal-dielectric configuration for the LRSPP waveguide, in which the fluorescent conjugated polymers MDMO-poly(phenylene vinylene) blended with the nonfluorescent poly(spirofluorene) (PSF) were used as the gain material [10]. In this structure, the metal layer is 4 nm thick Au. The authors replaced part of the PSF with a low-refractive-index compound to index-match the gain material with that of the dielectric cladding layers. Using this method, the authors could greatly reduce the propagation loss of the LRSPP mode, as shown in Fig. 7.1. A maximum gain of ~59 cm^{-1} at wavelength 600 nm is realized in the experiment.

Almost at the same time, De Leon et al. reported another method to index-match the refractive index of the gain material [11]. The authors fabricated an Au stripe of 20 nm thick on top of the SiO$_2$ and covered the structure with a gain layer of ~100 µm thick IR140 dye solution, as shown in Fig. 7.2. The solvent they used was a mixture of 30.4% ethylene glycol and 69.6% dimethyl sulphoxide. At room temperature, the dye solution was index-matched to SiO$_2$ at the emission wavelength, forming a symmetric LRSPP structure. The authors claimed a gain coefficient of 8.85 dBmm^{-1} in their structure.

These experiments demonstrated that the gain compensation for the propagation loss of LRSPPs can be achieved by highly fluorescent dye molecules or semiconducting conjugated polymers as long as a proper index-matching technique is implemented. The gain coefficients shown in these papers are below 100 cm^{-1}, still far below what is needed for short-range SPPs.

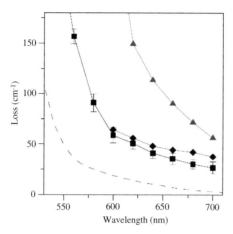

Figure 7.1 Measurement and optimization of the propagation loss in 4 nm thick gold LRSPP waveguides. Propagation loss (as derived from the exponential decay of scattered light intensity) for different plasmonic waveguides as a function of the wavelength for a structure without a gain material (squares), with a gain layer that was partly index-matched (triangles, nominal $\Delta n = 0.03$) and fully index-matched (diamonds). Error bars represent uncertainty of the fit to the exponential decay and are representative for all data sets. The dashed line represents a theoretical loss estimate (neglecting scattering) for the structure without a gain material. Reprinted by permission from Macmillan Publishers Ltd: [*Nature Photonics*] (Ref. [10]), copyright (2010).

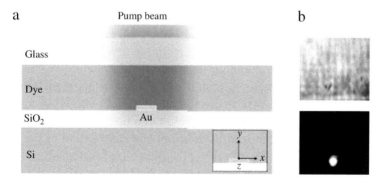

Figure 7.2 Guiding structure. (a) Cross-sectional view of the active structure (not to scale). The gain medium is in the form of a laser dye in a solution. Inset: coordinate system with the +z direction coming out of the page. (b) x-Polarized (top) and y-polarized (bottom) light collected from the output facet of the structure while being pumped by x-polarized light. Reprinted by permission from Macmillan Publishers Ltd: [*Nature Photonics*] (Ref. [11]), copyright (2010).

7.3 Stimulated Emission from Localized Surface Plasmon Resonance with a Gain Material

Due to its potential application in plasmon-based molecular sensors, the enhancement of localized SP resonance with a gain material has gained increasing attention in the plasmon community. The estimation for the critical gain coefficient has been worked out in a few theoretical papers, focusing on different geometries of the nanoparticles [17, 18]. They all start from the calculation of the polarizability of a nanoparticle in an applied electric field. One approximation that people usually make to simplify the calculation is the electrostatic approach, which is valid when the size of the metallic nanoparticle is much smaller than the wavelength of the radiation light. In this case, the electric potential Φ becomes the solution of Laplace equation $\nabla^2 \Phi = 0$, from which we can find the electric field distribution as $\mathbf{E} = -\nabla\Phi$. The detailed derivation can be found from Refs. [1, 2]. A more general form of the polarizability α of an isolated nanoparticle in the electric field is given by $\alpha = (4\pi)^{-1}(\varepsilon_m - \varepsilon_d)/[\varepsilon_d + p(\varepsilon_m - \varepsilon_d)]$, where ε_m and ε_d are respectively the complex permittivities of the metal and surrounding dielectric and p is the depolarization factor associated with the shape of the nanoparticle. In the case of a spherical nanoparticle, $p = 1/3$, and we get $\alpha \propto (\varepsilon_m - \varepsilon_d)/(2\varepsilon_d + \varepsilon_m)$. This result for a spherical particle can also be found from the classical book of Jackson [19]. We can see clearly that the resonance condition of the polarizability is satisfied when the value of the denominator $|\varepsilon_d + p(\varepsilon_m - \varepsilon_d)|$ is a minimum. In the case of small or slowly varying $\text{Im}(\varepsilon_m)$, the minimum is achieved when $\text{Re}(\varepsilon_m) = -(1-p)\text{Re}(\varepsilon_d)/p$. The resonance is limited by the nonvanishing term $\text{Im}(\varepsilon_m)$. However, when the passive dielectric material is replaced by a gain medium, the $\text{Im}(\varepsilon_d)$ is no longer zero but a negative number while being pumped. It can lead to the complete cancellation of the denominator term when $\text{Im}(\varepsilon_m) = -(1-p)\text{Im}(\varepsilon_d)/p$ is satisfied, and the resonance will be strongly enhanced and only be limited by the gain saturation. The critical gain coefficient needed to make the propagation loss zero is found to be $\gamma = (2\pi/\lambda_0)\text{Im}(\varepsilon_d)/(\text{Re}[\varepsilon_d])^{1/2}$, where λ_0 is the vacuum wavelength at the resonance frequency. If we use the Drude formula to approximate the dielectric function of the metal $\varepsilon_m = 1 - \omega_p^2/[\omega(\omega + i\Gamma)]$, the gain coefficient $\gamma = (2\pi/n\lambda_0)(\Gamma/\omega_p)[p/$

$(1-p)][1 + n^2(1-p)/p]^{3/2}$. For spherical nanoparticles, $\gamma = (\pi/n\lambda_0)(\Gamma/\omega_p)(1 + 2n^2)^{3/2}$. If we use $\lambda_0 = 630$ nm and $n = 1.33$ and Γ and ω_p are typical values for gold and silver, we can estimate $\gamma = \sim 10^3$ cm^{-1}.

This critical gain value is achievable with highly fluorescent materials with a stimulated cross section of approximately 5×10^{-16} cm^2, such as some conjugated polymers, dye molecules, or quantum dots (QDs). If we assume each excited molecule adsorbed on a nanoparticle a few nanometers in diameter, it gives the gain coefficient at the order of magnitude needed to compensate for the metallic loss at resonance.

Works regarding the increase of fluorescence intensity for dye molecules adsorbed on islands and films of nanoparticles whose plasmon resonance matches with the absorption band of the molecules were reported by Glass et al. and Ritchie and Burstein in the early 1980s [20–22]. The increase of absorption of molecules and QDs and the enhancement of their fluorescence intensity and modification to their emission spectra due to the presence of metallic nanostructures have been investigated intensively in recent years. Studies showed that if metallic nanoparticles were dispersed into an optical cavity, the pump power that is used to reach the threshold for lasing can be greatly decreased [23–25].

In Ref. [25], the authors discussed in detail the influence of Ag nanoparticle aggregates on various properties of a rhodamine 6G (R6G) solution. They studied the absorption, Rayleigh scattering, spontaneous emission, and stimulated emission from the R6G and Ag nanoparticle mixture separately. Their results showed that the increased scattering in the solution is due to an enhancement in the Rayleigh scattering cross section of the metallic particles in the presence of gain media. This scattering intensity can increase or decrease with different pump power values, which is explained with a theoretical model developed by Lawandy [26]. The increased spontaneous emission intensity from R6G is due to the enhanced absorption caused by the field enhancement associated with the SP resonances of Ag nanoparticles. The enhanced stimulated emission, which has the same nature as the enhanced absorption, is observed in both pump-probe and laser experiments. The most interesting result of their experiment is that in the presence of Ag aggregates, the value of amplification is not limited by the saturation-level characteristic of a pure dye solution but grows to a higher magnitude (Fig. 7.3).

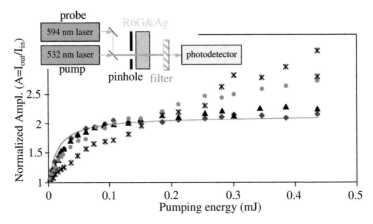

Figure 7.3 Amplification $A = I_{out}/I_{in}$ (at λ = 594 nm) as a function of the 532 nm pumping energy (after a 0.5 mm pinhole) measured in a series of R6G dye–Ag aggregate solutions in a 10 mm cuvette. Solid line: calculation corresponding to a pure dye, 1.25×10^{-5} M. Characters: experiment. Diamonds: pure dye, 1.25×10^{-5} M; triangles: dye, 1.25×10^{-5} M, Ag aggregate, 3.6×10^{11} particles/cm^3; circles: dye, 9.0×10^{-6} M, Ag aggregate, 2.5×10^{12} particles/cm^3; crosses: dye, 5.1×10^{-6} M, Ag aggregate, 5.2×10^{12} particles/cm^3. All amplification signals are normalized to the transmission in the corresponding not-pumped media. Inset: schematic of the experiment. Reprinted with permission from Ref. [25]. Copyright (2006) by the American Physical Society.

The concept of surface plasmon amplification by stimulated emission of radiation (SPASER) in a nanosystem was first explicitly predicted by Bergman and Stockman in 2003 [18]. In this paper, they stated that the radiation of SPASER is mostly composed of SPs, which undergo stimulated emission as photons, only that they can be confined to nanoscale. In SPASER, the active medium undergoes population inversion upon strong pump and transfers its excited energy to a resonant nanosystem via nonradiative transitions. The energy transfer from an excited molecule to a resonant SP oscillation in a metallic nanostructure can also be explained by Förster dipole–dipole interaction, which can also give rise to the main features of SPASER [27, 28].

SPASER was first demonstrated experimentally by Noginov et al. in 2009 [12]. The authors synthesized high-brightness luminescent core-shell nanoparticles, which are composed of a gold core, providing the plasmon modes, surrounded by a silica shell containing the Oregon Green 488 (OG-488) dye molecules, as the gain medium (shown in Fig. 7.4).

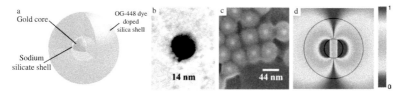

Figure 7.4 (a) Diagram of the hybrid nanoparticle architecture (not to scale), indicating dye molecules throughout the silica shell. (b) Transmission electron microscope image of Au core. (c) Scanning electron microscope image of Au/silica/dye core-shell nanoparticles. (d) SPASER mode (in a false color), with λ = 525 nm and Q = 14.8; the inner and outer circles represent the 14 nm core and the 44 nm shell, respectively. The field strength color scheme is shown on the right. Reprinted by permission from Macmillan Publishers Ltd: [*Nature*] (Ref. [12]), copyright (2009).

The diameter of the Au core is ~14 nm and that of the silica shell ~15 nm. There are roughly 2.7×10^3 molecules per nanoparticle. These nanoparticles were suspended in a water solution after cleaning, with a concentration equal to 3×10^{11} cm^{-3}. In this configuration, the extinction spectrum of the nanoparticle in the water solution overlaps with both the excitation and emission bands of the dye molecules incorporated in the silica shell. At a stronger pump intensity, the emission spectrum of the solution undergoes a characteristic peak narrowing, corresponding to the commencement of stimulated emission. The authors then went on showing that the full-width at half-maximum (FWHM) of the stimulated emission peak and the emission intensity changed with the pump energy following the same trend as a regular laser (shown in Fig. 7.5). To ensure that this observation is not caused by random lasing within a solution or a feedback provided by the cuvette walls, where the solution was contained, the authors diluted the solution by 100-fold and still observed the same spectral line without diminishing the ratio of stimulated emission intensity to spontaneous emission background. The authors then concluded that the emission spectra are characteristic of each nanoparticle.

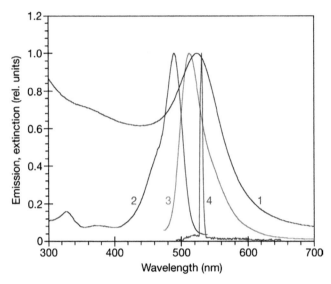

Figure 7.5 Normalized extinction (1), excitation (2), spontaneous emission (3), and stimulated emission (4) spectra of Au/silica/dye nanoparticles. The peak extinction cross section of the nanoparticles is 1.1×10^{-12} cm^2. The emission and excitation spectra were measured in a spectrofluorometer at low fluency. Reprinted by permission from Macmillan Publishers Ltd: [*Nature*] (Ref. [12]), copyright (2009).

The authors then estimated the number of excited molecules per nanoparticles that is needed to compensate for the loss in the localized SP resonance, which is about 2×10^3 molecules per nanoparticle, following a similar method as presented at the beginning of this section. This number is less than the available number of molecules that are incorporated in the silica shell. The pump energy that the authors used to achieve lasing is above that needed to saturate all OG-488 molecules. Therefore, it is assumed that all molecules within each nanoparticle are in their excited states. The authors then finally concluded that they have demonstrated a SPASER "nanolaser that is realized by each individual nanoparticle, making it the smallest reported in the literature and the only one to date operating in the visible range."

This result has quite a significant impact on the plasmon community. Its demonstration means that it is possible to have a coherent light source whose size is comparable with the typical size of a function unit in the current complementary metal-oxide-semiconductor (CMOS) industry and promises a bright future for plasmon-based integrated photonics.

Before finishing up this section, there is one more experiment we would like to mention. In 2010, Walther et al. demonstrated an electrically pumped microcavity terahertz laser based on an LC circuit resonant oscillation with a mode volume at a deep subwavelength scale (Fig. 7.6) [29]. The Au metal cap on top of the active medium (quantum cascade laser) makes it possible to confine the electric field to the gain region. The simulation of electric field distribution shows a localized resonance at the active medium. They estimated that the spontaneous emission life time of the gain medium is enhanced 17-fold. The quality factor they obtained in the experiment is from 11 to 21, a value that is limited by the metallic loss. On the basis of the same working principles, they speculated that similar lasers can be realized at higher frequencies and even into the near-infrared range.

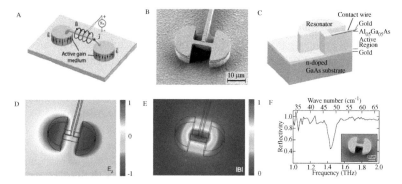

Figure 7.6 (A) Schematic of the LC laser. **J** is the alternating current in the resonator, **B** is the induced magnetic field, and **E** is the electric field. The active gain medium is biased by the voltage source V_{DC}. (B) Scanning electron micrograph of the LC laser device. (C) Schematic cross section through the device along the symmetry axis. The red layer is undoped $Al_{0.5}Ga_{0.5}As$ and prevents current injection into the active region below the bonding pad. (D, E) Finite element simulations of the electromagnetic field in the resonator showing the dominating electric field component E_z and the norm of the magnetic field $|\mathbf{B}|$. (F) Measured reflectivity at 10 K of an array of 400 identical LC resonators, shown in the inset and designed for a frequency of 1.45 THz, without a gain medium and without an electrical connection. From Ref. [29]. Reprinted with permission from AAAS.

7.4 Gain-Assisted Hybrid Surface Plasmon Propagation: Lasing and Amplification

7.4.1 Hybrid Surface Plasmon Lasers

In parallel with the works presented in last section, people also tried to increase the propagation length of propagating SPPs, which may serve as optical signal carriers with a subwavelength mode confinement, with gain materials. In this section, we will discuss the lasing and amplification of a lossy propagating SPP mode.

For localized SP resonance, the compensation of loss by the gain material is shown as the enhanced polarizability and narrowing of resonance peak line width. For propagating SPPs, people first studied the enhanced propagation length upon the implementation of gain media with various waveguide configurations. For nanoparticle chains embedded in a gain medium, Citrin et al. showed that the increase in interparticle coupling strength caused by the active material can lead to an extended propagation distance [30]. Avrutsky theoretically investigated SPPs at the metal–dielectric interface with a strong optical gain [31]. Interestingly, he predicted that at the resonance condition, the effective refractive index of the SPPs can go to infinity, which is only limited by the gain saturation, and results in strong localization of the surface wave at the interface [31]. This effect can be simply demonstrated with the dispersion relation of SPPs existing at the metal–dielectric interface $n_{eff} = \sqrt{\frac{\varepsilon_m \varepsilon_d}{\varepsilon_m + \varepsilon_d}}$. As we discussed in the localized SP resonance case, if the dielectric is carefully chosen so that $Re(\varepsilon_d) = Re(\varepsilon_m)$, the resonance condition will be fulfilled. With optical gain in the same dielectric, the imaginary part of the two materials can also be cancelled, meaning $Im(\varepsilon_d) = Im(\varepsilon_m)$. Then n_{eff} becomes infinitely large, making the SP mode extremely localized at the interface. However, this phenomenon is yet to be demonstrated by the experiment. For materials not satisfying the resonance condition, the gain coefficient is written as $\gamma = k_0 Im(\varepsilon_d)/\sqrt{Re(\varepsilon_d)}$, where k_0 is the wave number of light in vacuum. For semiconductor materials with $n = 3.4$, γ is expected to be ~4590 cm^{-1} for lossless propagation at the semiconductor–Ag interface with a wavelength of 650 nm, even higher than that of a localized SP resonance case and just falling within the limits that can be offered by semiconductor quantum wells and QDs. However, for lower-refractive-index materials such as doped glass or conjugated polymers, this requirement can be lowered to <1000 cm^{-1} [32].

The amplification of the SPPs at the interface of a silver film and a pumped dye solution was first demonstrated by Seidel et al. in 2005, with an observed increase in the metal reflectivity by 0.001% [33].

To maintain the deep subwavelength confinement offered by the short-range SPPs and be able to further decrease the propagation loss in the SPP mode, Oulton et al. proposed theoretically a hybrid plasmonic mode, which exists at the metal–insulator–semiconductor interfaces [34]. This hybrid plasmonic waveguide was then demonstrated by Oulton et al. later in 2009, showing mode areas as small as $\lambda^2/400$ [13]. In this paper, the authors also demonstrated the full compensation of propagation loss of this hybrid mode by the realization of SPASER at 489 nm at a low temperature of <10K.

This nanometer-scale plasmonic laser is composed of a single crystal cadmium sulphide (CdS) nanowire on a silver film, between which is the gap layer of magnesium fluoride (MgF$_2$), as shown in Fig. 7.7.

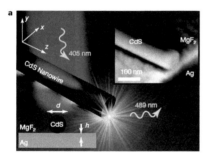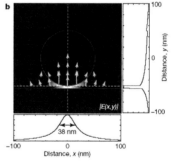

Figure 7.7 (a) The plasmonic laser consists of a CdS semiconductor nanowire on top of a silver substrate, separated by a nanometer-scale MgF$_2$ layer of thickness h. This structure supports a new type of plasmonic mode, the mode size of which can be a hundred times smaller than a diffraction-limited spot. The inset shows a scanning electron microscope image of a typical plasmonic laser, which has been sliced perpendicular to the nanowire's axis to show the underlying layers. (b) The simulated electric field distribution and direction $|E(x, y)|$ of a hybrid plasmonic mode at a wavelength of $\lambda = 489$ nm, corresponding to the CdS I$_2$ exciton line. The cross-sectional field plots (along the broken lines in the field map) illustrate the strong overall confinement in the gap region between the nanowire and the metal surface, with a sufficient modal overlap in the semiconductor to facilitate gain. Reprinted by permission from Macmillan Publishers Ltd: [*Nature*] (Ref. [13]), copyright (2009).

To illustrate the unique properties of a hybrid plasmonic mode, the authors compared the plasmon lasers with conventional photonic lasers, which consist of CdS nanowires on a quartz substrate, as shown in Fig. 7.8. As there is literally no cutoff frequency for the hybrid plasmonic waveguide mode, it was shown clearly that for the plasmon lasers, the lasing threshold is only weakly dependent on the diameter of the nanowire, while for photonic lasers the lasing threshold is strongly dependent on the diameter of the nanowire, owing to the reduction in the total gain volume for smaller nanowires.

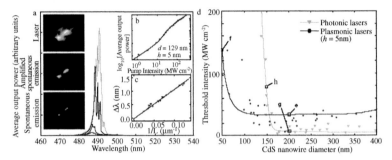

Figure 7.8 Laser oscillation and threshold characteristics of plasmonic and photonic lasers. (a) Laser oscillation of a plasmonic laser, d = 129 nm and h = 55 nm (longitudinal modes). The four spectra for different peak pump intensities exemplify the transition from spontaneous emission (21.25 MW cm^{-2}) via amplified spontaneous emission (32.50 MW cm^{-2}) to full laser oscillation (76.25 MW cm^{-2} and 131.25 MW cm^{-2}). (b) The nonlinear response of the output power to the peak pump intensity. The relationship between mode spacing $\Delta\lambda$ and nanowire length L in (c) indicates a high group index of 11 due to the high material gain. The pictures on the left correspond to microscope images of a plasmon laser with d = 66 nm, exhibiting spontaneous emission, amplified spontaneous emission, and laser oscillation, where the scattered light output is from the end facets. (d) Threshold intensity of plasmonic and photonic lasers versus the nanowire diameter. The experimental data points correspond to the onset of amplified spontaneous emission, which occurs at slightly lower peak pump intensities compared to the threshold of gain saturation. Amplified spontaneous emission in hybrid plasmonic modes occurs at moderate pump intensities of 10–60 MW cm^{-2} across a broad range of diameters. Reprinted by permission from Macmillan Publishers Ltd: [Nature] (Ref. [13]), copyright (2009).

The authors found a strong increase of spontaneous emission rate, which was associated with Purcell effect of the strong confinement of the plasmon mode. A Purcell factor larger than 5 was achieved with an insulator layer of 5 nm thickness. The authors also estimated the saturated internal gain of the laser from the measurements on averaged group index, and the gain coefficient is over 10,000 cm^{-1}, enough to compensate for the propagation loss of the mode, which is around 4100 cm^{-1}, and transmission loss at the end facets of the nanowires.

In 2011 Ma et al. from the same group demonstrated a plasmonic laser operating at room temperature with a subwavelength mode volume [14]. This laser has the same three components in a vertical direction as the previous one. However, it adopted a squared top layer, which made it benefit from a much-decreased cavity loss, as shown in Fig. 7.9. When the SPP mode is hybridized with the transverse magnetic (TM) mode of CdS square, the effective refractive index increases dramatically, resulting in total internal reflection as it reflects off the edge of the square top. For the photonic mode, on the other hand, its profile becomes increasingly delocalized as the thickness of MgF_2 decreases and cannot undergo total internal reflection. Therefore, for a CdS layer of 45 nm and a MgF_2 layer of 5 nm, only the plasmonic mode starts to lase as the pump power increases.

The authors also went on showing that the number of lasing cavity modes can be decreased to one by tailoring the shape of the top layer. In this experiment, the Purcell factor is estimated to be 17, with the combination effect of strong mode confinement and improved cavity. The SPP loss is estimated to be 6323 cm^{-1}, much smaller than the gain that can be provided by the semiconductor material, which is around 10,000 cm^{-1}.

Since the publication of the above two papers, hybrid plasmon modes at the semiconductor–insulator–metal interfaces have become a popular model system to achieve SPASER with a well-confined mode volume. Exploiting a triangular cross section of the GaN nanowire, Zhang et al. achieved a hybrid plasmonic laser at an ultraviolet wavelength (370 nm) and with a low lasing threshold at room temperature [35]. Lu et al. adapted epitaxial-grown Ag as the substrate in an effort to further decrease the lasing threshold of SPASER [15]. They demonstrated a low-threshold hybrid plasmonic laser pumped by a continuous-wave (CW) laser at 120 K. In 2014, the same authors reported ultralow-threshold CW hybrid plasmonic lasers operating at three primary colors, with a lasing threshold as low as 10 W/cm^2 [36].

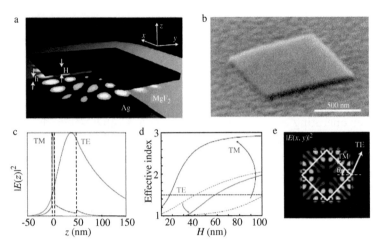

Figure 7.9 The room-temperature plasmon laser. (a) Schematic diagram of the room-temperature plasmon laser showing a thin CdS square atop a silver substrate separated by a 5 nm MgF_2 gap, where the most intense electric fields of the device reside. (b) Scanning electron micrograph of the 45 nm thick, 1 μm length CdS square plasmon laser studied here. (c) The electric field intensity distribution of the two modes of the system along the z direction. Although TM modes are localized in the gap layer, transverse electric (TE) modes are delocalized from the metal surface. (d) The effective index of TM and TE waves with (solid line) and without (dashed line) the metal substrate. TM waves strongly hybridize with SPPs, resulting in strong confinement within the gap region (TM line in (c)) accompanied by a dramatic increase in momentum (solid TM line in (d)) with respect to TM waves of the CdS square alone (dashed TM line in (d)). However, the delocalized TE waves (TE line in (c)) show decreased momentum (solid TE line in (d)) with respect to TE waves of the CdS square alone (dashed TE line in (d)). (e) Electric field intensity distribution of a TM mode in the x and y directions. Although both mode polarizations are free to propagate in the plane, only TM modes have a sufficiently large mode index to undergo efficient total internal reflection, providing the feedback for lasing. Reprinted by permission from Macmillan Publishers Ltd: [*Nature Materials*] (Ref. [14]), copyright (2011).

To demonstrate the feasibility of large-scale integrated plasmonic lasers at room temperature, Liu et al. used industry-relevant heterogeneous quaternary AlGaInP as the gain material and top-down patterning and etching techniques to demonstrate room-temperature, low-threshold, red-emitting hybrid plasmonic lasers of various cavity shapes, including waveguides, rings, circular disks, and rectangles of a mode volume down to $0.2(\lambda/2n_{\text{eff}})^3$ with footprints as small as 350 nm × 390 nm [37]. The authors observed a strongly enhanced spontaneous emission rate with a Purcell factor as large as 29. In particular, the authors attributed the enhanced stimulated cross section observed in the hybrid plasmonic lasers to the mode confinement (Purcell factor). This connection is critical for achieving low lasing threshold at extremely small hybrid plasmonic cavities.

An interesting twist to the vertical layout of semiconductor-insulator-metal configuration was reported by Wu et. al in 2013 [38]. In this paper, the authors coupled a CdSe nanowire with a silver nanowire through near-field contact, so the entire structure forms an X shape. Different from the hybrid plasmonic modes, where the photonic and plasmonic components are hybridized in the transverse direction, here the photon cavity mode (within a CdSe nanowire) and the plasmon cavity mode (within a Ag nanowire) are hybridized along the longitudinal direction and thus spatially separated, as shown in Fig. 7.10. By pumping on the CdSe nanowire portion of the hybrid cavity, the authors can get lasing output from the end facet of the Ag nanowire, with a mode area of 0.008 λ^2.

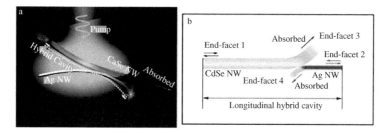

Figure 7.10 Schematic illustration of a hybrid photon-plasmon nanowire laser. (a) Closed-up view of the coupling area indicating the coupled hybrid cavity (marked by the dashed line), which serves as the hybrid photon-plasmon lasing cavity. (b) Cavity formation in the X structure. Reprinted with permission from Ref. [38]. Copyright (2013) American Chemical Society.

In 2011, Kitur et al. demonstrated an SPP microcylinder cavity laser (Fig. 7.11). The microcylinder consists of Au or Ag microwires (MWs), whose diameters are 10 and 27.2 µm, respectively [39]. To make the lasers, the authors dipped the MWs into polymethyl methacrylate (PMMA) and R6G solution and then pulled them out and let them dry in air. This process eventually leaves a dried dye-doped PMMA film outside the MWs, and the quality of the film depends on the thickness of the PMMA layer. As a reference, the authors also fabricated the dye-doped PMMA layer outside a fiber glass. The authors observed lasing of multiple whispering gallery modes from the metal MWs up on an optical pump and concluded that these modes are from the propagating SPPs at the inner interface of metal and PMMA film from the spectral spacing of the lasing peaks.

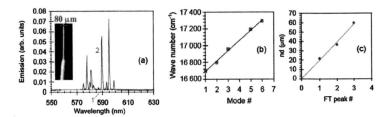

Figure 7.11 (a) Emission spectrum of R6G/PMMA-coated gold wire; trace 1 is with 0.026 mJ and trace 2 is with 0.052 mJ. Inset: microphotograph of the wire; (top part) 10 μm bare gold wire and (bottom part) ~30 μm wire coated with R6G/PMMA. (b) Dependence of the mode energy versus the mode number. (c) Position of the Fourier transform (FT) peak (in units of nd) versus the FT peak number. Reprinted with permission from Ref. [39]. Copyright (2011) by the American Physical Society.

7.4.2 Amplification of Hybrid Surface Plasmon Polaritons

Despite the tremendous successes of hybrid plasmonic lasers, they can only provide nanoscale light sources. A successful signal transport within the plasmonic circuits requires in situ amplification mechanism to compensate for the signal attenuation due to propagation loss, so the information carried by the plasmonic signal can be transferred from one part of the circuit to another. This is crucial to the scalability and cascadability of the plasmonic circuits.

We experimentally demonstrated in 2013 that a weak hybrid plasmonic signal can be amplified and fully loss compensated by the optical pump-probe technique. The setup to execute the signal amplification through a pump probe is demonstrated in Fig. 7.12. In the experiments, we measured the loss and gain coefficients of a hybrid plasmonic waveguide, consisting of a CdSe nanobelt (NB) separated from a silver surface by an Al_2O_3 thin layer, using the pump-probe technique [16]. We demonstrated that these hybrid plasmonic waveguides support an ultrahigh optical gain, with a gain coefficient exceeding 6755 cm^{-1}.

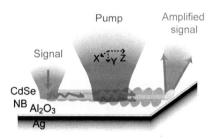

Figure 7.12 Diagram of CdSe NB/Al$_2$O$_3$/Ag hybrid plasmonic waveguide and the excitation and amplification of the input probe signal when operated in a pump-probe setup [16].

To characterize how the pump-probe technique can adapt to different polarizations of the probe signal, we alter the polarization of the input signal. This in turn changes the polarization of the waveguided hybrid SP modes. By measuring the output intensity with and without the pump light, we can directly extract the gain coefficient of each hybrid SP mode. Figure 7.13 shows the signal amplification through stimulated emission on probe signals of orthogonal polarizations. It is quite clear that the (attenuated) probe signals of both polarizations are significantly amplified when the pump light is on. Taking into account the propagation loss of a waveguided signal, we concluded that 99% of the propagation loss of a TM-dominant (input beam polarized along the long axis of the NB) hybrid plasmonic mode is compensated, while only 58% of that of the higher-order mode (input beam polarized perpendicular to the long axis of the NB) is compensated due to the stronger loss in the higher-order mode. These gain coefficients are much higher than the ones observed in Refs. [10, 11] but similar to the values calculated in Refs. [13, 14].

More importantly, we show that the loss compensation works in a relatively broad spectral band and for input signals of tunable polarizations, in contrast to nanolasing that requires single-mode operation. Our result is an important step forward to the realization of integrated plasmonic circuits.

From the experiments discussed in this section, we can see clearly that by carefully engineering the gain materials at the close vicinity of metal films or metal wires, the propagation loss of short-range SPPs caused by the damping of metal can be fully compensated by the gain media, evident by the lasing of SPASER under optical pumping. On the basis of the same principles, amplifiers that can be used to magnify a weak short-range SPP signal with the similar gain coefficients as that for the SPASER have also been demonstrated experimentally. These plasmonic lasers and amplifiers can greatly advance the development of on-chip integrated plasmonic optics.

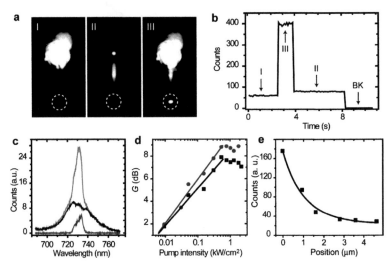

Figure 7.13 (a) Optical images obtained with a 730 ± 5 nm band pass filter, corresponding to a probe signal launched from the top end of the hybrid plasmonic waveguide (NB 177 nm × 140 nm × 8.6 µm) and emitted from the bottom end, highlighted by the dashed circle (I), photoluminescence (PL) with the pump only (II), and the amplification of the probe signal when both pump and probe are present (III). The two ends of the waveguide are clearly visible as the top and bottom bright spots in (II). (b) Time trace of the output intensity at the emission end of the hybrid plasmonic waveguide. The dashed circles in (a) mark the area where the time trace was obtained. BK indicates the background dark counts. (c) Typical set of emission spectra, representing the probe signal (bottom), PL (middle), and total output (top). The pump intensity is 0.77 kW/cm^2. (d) Gain G (dB) versus pump intensity. The symbols ■ and ● correspond to the measured gains of the input laser beam polarized along and perpendicular to the waveguide, respectively, and the solid lines correspond to logarithmic fits to the G value at a pump intensity lower than 0.6 kW/cm^2. (e) Propagation loss measurement. Output light intensity as a function of pump position and fit to linear combination of two exponential decay functions [16].

7.5 Summary and Future Perspective

In summary, we have reviewed the recent progress on SP-based lasers and amplifiers. Great successes have been achieved in this field. Gain coefficients of the order of 100 cm^{-1} were realized on the amplifier on the basis of long-range propagating SPs. High fluorescent dye molecules and semiconductor conjugated polymers were used as gain materials. These types of gain materials can also be implemented into the metal-oxide core-shell structures. Stimulated emission from strongly enhanced localized SP resonances was observed in the latter case. The gain coefficient in this case was estimated to be in the range of 1000 cm^{-1}. The highest gain coefficient, 10,000 cm^{-1}, has been demonstrated by semiconductor materials in the hybrid SP lasers using a semiconductor-insulator-metal structure.

The demonstration of amplifiers based on hybrid SPPs is critical for further development of integrated plasmonic circuits. Our experiments also show a 35-fold decrease (compared to the photonic case) in the transparency threshold, at which population inversion starts. This means that the gain material in hybrid plasmonic configuration reaches population inversion at a low pump intensity. We attribute this to the "hot" carrier transfer at the metal–insulator–semiconductor interfaces upon the radiation of pump light. Combining this concept with enhanced stimulated emission due to the Purcell effect, gain-assisted plamonic propagation and cavities promise high-gain, low-threshold plasmonic amplifiers and lasers in the future.

The demonstration of SPASER and plasmon-based amplifiers has given solutions to one of the largest obstacles that hinder the further advancement of plasmonic devices, propagation loss, and has opened new possibilities to the development of fully integrated nanoplasmonic circuits in the future.

References

1. S. A. Maier, *Plasmonics: Fundamentals and Applications* (Springer, New York, 2007).
2. V. M. Shalaev and S. Kawata, *Advances in Nano-Optics and Nano-Photonics* (Elsevier, Amsterdam, 2007).

3. Y. R. Fang, Z. P. Li, Y. Z. Huang, S. P. Zhang, P. Nordlander, N. J. Halas, and H. X. Xu, *Nano Lett.*, **10**, 1950–1954 (2010).
4. A. V. Akimov, A. Mukherjee, C. L. Yu, D. E. Chang, A. S. Zibrov, P. R. Hemmer, H. Park, and M. D. Lukin, *Nature*, **450**, 402–406 (2007).
5. H. Wei, Z. P. Li, X. R. Tian, Z. X. Wang, F. A. Cong, N. Liu, S. P. Zhang, P. Nordlander, N. J. Halas, and H. X. Xu, *Nano Lett.*, **11**, 471–475 (2011).
6. H. Wei, Z. X. Wang, X. R. Tian, M. Käll, and H. X. Xu, *Nat. Commun.*, **2**, 387 (2011).
7. P. Berini and I. De Leon, *Nat. Photonics*, **6**, 16–24 (2012).
8. M. C. Gather, *Nat. Photonics*, **6**, 708 (2012).
9. R. M. Ma, R. F. Oulton, V. J. Sorger, and X. Zhang, *Laser Photonics Rev.*, **7**, 1–21 (2013).
10. M. C. Gather, K. Meerholz, N. Danz, and K. Leosson, *Nat. Photonics*, **4**, 457–461 (2010).
11. I. De Leon and P. Berini, *Nat. Photonics*, **4**, 382–387 (2010).
12. M. A. Noginov, A. M. Belgrave, R. Bakker, V. M. Shalaev, E. E. Narimanov, S. Stout, E. Herz, T. Sutewong, and U. Wiesner, *Nature*, **460**, 1110–1112 (2009).
13. R. F. Oulton, V. J. Sorger, T. Zentgraf, R.-M. Ma, C. Gladden, L. Dai, G. Bartal, and X. Zhang, *Nature*, **461**, 629–632 (2009).
14. R. M. Ma, R. F. Oulton, V. J. Sorger, G. Bartal, and X. Zhang, *Nat. Mater.*, **10**, 110–113 (2011).
15. Y. J. Lu, J. Kim, H. Y. Chen, C. H. Wu, N. Dabidian, C. E. Sanders, C. Y. Wang, M. Y. Lu, B. H. Li, X. G. Qiu, W. H. Chang, L. J. Chen, G. Shvets, S. Gwo, and C. K. Shih, *Science*, **337**, 450–453 (2012).
16. N. Liu, H. Wei, J. Li, Z. X. Wang, X. R. Tian, A. L. Pan, and H. X. Xu, *Sci. Rep.*, **3**, 1967 (2013).
17. N. M. Lawandy, *Appl. Phys. Lett.*, **85**, 5040–5042 (2004).
18. D. J. Bergman and M. I. Stockman, *Phys. Rev. Lett.*, **90**, 027402 (2003).
19. J. D. Jackson, *Classical Electrodynamics*, 3rd ed. (John Wiley & Sons, New York, 1999).
20. A. M. Glass, P. F. Liao, J. G. Bergman, and D. H. Olson, *Opt. Lett.*, **5**, 368–370 (1980).
21. A. M. Glass, A. Wokaun, J. P. Heritage, J. G. Bergman, P. F. Liao, and D. H. Olson, *Phys. Rev. B*, **24**, 4906–4909 (1981).
22. G. Ritchie and E. Burstein, *Phys. Rev. B*, **24**, 4843–4846 (1981).

23. W. Kim, V. P. Safonov, V. M. Shalaev, and R. L. Armstrong, *Phys. Rev. Lett.*, **82**, 4811–4814 (1999).
24. V. P. Drachev, W. T. Kim, V. P. Safonov, V. A. Podolskiy, N. S. Zakovryashin, E. N. Khaliulin, V. M. Shalaev, and R. L. Armstrong, *J. Mod. Opt.*, **49**, 645–662 (2002).
25. M. A. Noginov, G. Zhu, M. Bahouza, J. Adegoke, C. Small, C. Darvison, V. P. Drachev, P. Nyga, and V. M. Shalaev, *Phys. Rev. B*, **74**, 184203 (2006).
26. N. M. Lawandy, *Proc. SPIE*, **5924**, 59240G (2005).
27. T. Förster, *Ann. Phys.*, **2**, 55–75 (1948).
28. D. L. Dexter, T. Förster, and R. S. Knox, *Phys. Stat. Solidi.*, **34**, K159–K162 (1969).
29. C. Walther, G. Scalari, M. I. Amanti, M. Beck, and J. Faist, *Science,* **327**, 1495–1497 (2010).
30. D. S. Citrin, *Opt. Lett.*, **31**, 98–100 (2005).
31. I. Avrutsky, *Phys. Rev. B*, **70**, 155416 (2004).
32. S. A. Maier, *Opt. Commun.*, **258**, 295–299 (2006).
33. J. Seidel, S. Grafström, and L. Eng, *Phys. Rev. Lett.*, **94**, 177401 (2005).
34. R. F. Oulton, V. J. Sorger, D. A. Genov, D. F. P. Pile, and X. Zhang, *Nat. Photonics*, **2**, 496–500 (2008).
35. Q. Zhang, G. Y. Li, X. F. Liu, F. Qian, Y. Li, T. C. Sum, C. M. Lieber, and Q. H. Xiong, *Nat. Commun.*, **5**, 4953 (2014)
36. Y. J. Lu, C. Y. Wang, J. Kim, H. Y. Chen, M. Y. Lu, Y. C. Chen, W. H. Chang, L. J. Chen, M. I. Stockman, C. K. Shih, and S. Gwo, *Nano Lett.*, **14**, 4381–4388 (2014).
37. N. Liu, A. Gocalinska, J. Justice, F. Gity, I. Povey, B. McCarthy, M. Pemble, E. Pelucchi, H. Wei, C. Silien, H. X. Xu, and B. Corbett, *Nano Lett.*, **16**, 7822–7828 (2016).
38. X. Q. Wu, Y. Xiao, C. Meng, X. N. Zhang, S. L. Yu, Y. P. Wang, C. X. Yang, X. Guo, C. Z. Ning, and L. M. Tong, *Nano Lett.*, **13**, 5654–5659 (2013).
39. J. K. Kitur, V. A. Podolskiy, and M. A. Noginov, *Phys. Rev. Lett.*, **106**, 183903 (2011).

Index

absorption, 2, 7–9, 12, 15, 29–31, 48, 66, 68, 85–86, 115, 117, 127, 204, 207
  interband, 7
  preferential, 1
  surface-enhanced infrared, 15
absorption band, 143, 207
absorption spectrometer, 29
addition theorems, 31–32, 35
additive phase accumulation, 97
admixture, 71–72
adsorption, 121
AFM, *see* atomic force microscopy
aggregation, 122, 132
  irreversible, 122
  spontaneous, 122
Ag nanoparticles, 61, 66, 76–77, 111–12, 125, 174, 178, 207
Ag nanospheres, 60, 66
Ag NW, 120, 140, 142, 144–48, 150, 153, 155–60, 163–65, 167–70, 174, 176, 178, 180, 186, 192–94, 217
  bare, 153–54
  branched, 182
  coupled, 193
  glass-supported, 153
  molecule-coated, 160
  straight, 142
Ag NW waveguide, 143, 160
Ag spheres, 38–39, 46, 62
Ag trimer antenna, 76
amplification, 202–3, 207–8, 212–13, 219–20
amplifiers, 220–21
  low-threshold plasmonic, 221
  plasmon-based, 221

amplitude, 14, 95, 146, 155
amplitude decays, 13
amplitude signals, 95
AND gate, 187
angular momentum, 21, 24, 58, 66–67, 120
  optical, 171
  orbital, 172
antennas, 56, 85, 89, 93, 95–97, 129
  coupled gold rod, 96
  gold, 89
  gold bar, 89
  optical, 140
  surface-plasmon-coupled, 86
  triangle, 90
antibonding, 71
antinodes, 177, 180, 191
anti-Stokes scattering, 56
approximation
  modified long-wavelength, 7, 118
  quasi-static, 6, 57
  time-dependent local density, 78
arrays, 73, 75, 77, 127, 211
  atom, 21
  nanogap, 75
  ordered, 73, 81
  ordered nanoparticle, 75
  patterned, 128
  uniform, 75
asymmetric dielectric environment, 164
asymmetric emission, 164
asymmetry, 96, 155
  structural, 181
atomic force microscope, 176

Index

atomic force microscopy (AFM), 91–92

Babinet–Soleil compensator, 185
band gap, 183
bending loss, 162–63
bending radius, 162–63
Bessel beam, 116
Bessel functions, 24–25, 30, 146
blue shift, 77
Bohr's radius, 79
Boltzmann constant, 111
Boolean logic gates, 189
boundary conditions, 22, 28, 33, 58
boundary element method, 46
bowties, 86, 94, 105
branched NW, 157, 184
bright-field optical images, 160
broadband nanosource, 171
Brownian motion, 122, 128–29

carbon nanotube, 141
Cartesian coordinates, 41
cascadability, 218
CCD, *see* charge-coupled device
CD, *see* circular dichroism
CDA, see coupled dipole approximation
charge-coupled device (CCD), 142
Cherenkov radiation, 168
chirality, 168, 172
  optical, 103
  right-handed, 155
circuits
  integrated, 137–38, 147, 194
  integrated nanoplasmonic, 221
circular dichroism (CD), 103–4
circular polarizations, 168, 170, 183–84
  opposite, 172–73, 182–84
classical electrodynamics, 16, 222
Clausius–Mossotti polarizability, 5

CMOS, *see* complementary metal-oxide-semiconductor
coefficient, 28, 36
  attenuation, 162
  critical gain, 202, 206
  damping, 3
  diffusion, 80–81
  expansion, 26, 28–29, 41
  gain, 204, 207, 212, 215, 218–21
  incidence, 41
  loss, 218
  Mie, 28, 41
  scattering, 32–33
  translation, 31–32, 35–36, 41–42
collimated SPP beams, 167–68
colloidal lithography, cheap, 98
complementary metal-oxide-semiconductor (CMOS), 211
configuration, 62, 68, 93, 163, 209
  dielectric-metal-dielectric, 204
  hybrid plasmonic, 221
  semiconductor-insulator-metal, 217
confinement, 69, 129, 213–14, 216
conjugated polymers, 204, 207, 213, 221
continuous wave (CW), 215
continuum, 72, 170
  wire plasmon, 71–72
conventional Mie theory, 23, 27, 31–32
coordinates, 31, 35–37, 114, 165
  azimuthal, 35
  cylindrical, 145
  radial, 80
  spherical, 23
Coulomb repulsion, 121, 124
counterbeams, 130
counterions, 122
coupled dipolar mode shifts, 66

coupled dipole approximation (CDA), 113
coupled nanoantennas, 93, 95–96, 105
coupled nanoparticles/nanowires, 132
coupling, 14, 16–17, 66, 68, 71–72, 75, 82, 93, 97, 102, 128–29, 140, 171, 176, 192–94
  coherent, 176
  directional, 173
  grating, 12
  interparticle, 57, 63, 124
  plasmon, 14–16
  plasmon-exciton, 174
  plasmonic, 97
  prism, 12
crystal, 21–22, 138, 160
  birefringence, 101
CW, *see* continuous wave

damped Lorentz oscillator, 2
damping, 3–4, 10, 12, 57, 159, 161, 220
  effective, 57
  low, 3
  nonradiative, 9–10
dark field (DF), 117–18, 120, 122–23, 125–26, 160, 163
dark-field scattering spectrum, 90
dark-field spectroscopy, 29
DDA, *see* discrete dipole approximation
decay, 13, 140, 146, 174, 176
  exciton, 173
  exponential, 205
  nonradiative, 10
demultiplexer, 139, 178
depolarization, 99–103, 206
depolarization ratio, 76, 99–100, 102

Derjaguin–Landau–Verwey–Overbeek (DLVO), 121–22, 124
devices, 144, 192, 202, 211, 216
  charge-coupled, 142
  electrical, 85
  optical, 104
  optical circuit, 132
DF, *see* dark field
DF scattering, 118, 121–22, 126
diamond nanocrystals, 176
dielectric, 4, 12, 110, 117, 139, 192, 194, 202–3, 206, 212
dielectric constant, 2, 5–6, 8, 47, 51, 64
dielectric environment, 147, 149, 184, 194
  homogeneous, 145, 164–65, 183
dielectric function, 3–4, 7–8, 39, 46, 49, 51, 206
dielectric layer, 144, 180, 204
dielectric material, 171, 206
dielectric medium, 10–11, 14, 171–72
dielectric permittivity, 22
dimerization, 123–25
  spontaneous, 112, 132
  spontaneous optical, 125
  stable, 122
dimers, 38–39, 60–68, 73–77, 79–80, 86, 98–102, 111–13, 118, 120–25, 128–29
  bimetallic, 97
  bimetallic nanodisk, 98
  gold nanodisk, 77
  hot, 61
  isolated, 74
  metal nanoshell, 73–74
  model nanosphere, 63
  nanocrystal, 98–99
  planar, 97
dipolar interaction, 59, 74
dipolar nanoparticle plasmons, 71

dipolar resonance, 6, 91–92
dipolar SPR, 87
dipole–dipole interaction, 113, 208
dipole emission, 76, 102–3, 164
dipole mode, 90, 116
dipole plasmon resonance, 92
Dirac matrices, 171
Dirac notation, 25
discrete dipole approximation
    (DDA), 2, 46–47, 58–59, 92,
    111, 119
dispersion relation, 11, 71, 153–54
DLVO, see Derjaguin–Landau–
    Verwey–Overbeek
Drude dielectric function, 8
Drude expression, 4
Drude formula, 206
Drude model, 3–4, 6, 12, 51, 57
Drude permittivity, 11

EBL, see electron beam lithography
eigenenergies, 2
eigenfunction set, 24
eigenmodes, 145–46, 149
electric field, 5–6, 8, 24–26, 39–40,
    45–47, 50, 56–59, 72, 89,
    117–19, 146, 148–49, 163–65,
    170, 211
  confined, 89
  external, 63, 71
  induced, 9
  instantaneous, 147
  static, 5–6
  time-varying, 6
electric field distribution, 59, 100,
    143, 146–48, 172–73, 177,
    181–82, 206, 211
electric field intensity, 88, 157,
    182, 188, 191, 216
  local, 29, 141, 143, 177, 191
electromagnetic (EM), 1, 22, 56,
    110

electromagnetic coupling, 162, 174
electromagnetic energy, 104,
    137–38, 156
electromagnetic field, 96, 145, 211
electromagnetic waves, 203
electromagnetism, 29
electron beam lithography (EBL),
    75
electron–hole pair, 9, 144–45
electronic coupling, 77–78
electronic excitation mode,
    collective, 56
electrons, 3, 6, 78–79, 82, 137, 145
  conduction, 1, 4
  excited, 10
  free, 2–3, 117
  high-energy, 141
  hot, 16
electrostatic repulsion, 121–22
ellipticity, 77
EM, see electromagnetic
EM coupling, 14, 16, 57, 62, 65, 68,
    71, 78, 102, 140, 176, 193, 217
EM field distribution, 29, 56,
    79–80, 82
EM field enhancement, 1, 64, 79
EM waves, 1, 6, 22–24, 27, 47, 58
emission, 77, 85–86, 96–99, 101,
    159–60, 163–65, 168–70, 173,
    177–78, 185, 193–94, 209–10,
    214, 220
  antenna, 76
  far-field, 96
  fluorescent, 160
  measured SERS, 77
  remote, 1
  single-photon, 140
emission intensity, 159–60,
    164–65, 169, 174, 177,
    179–80, 207, 209
  distance-dependent, 159, 161
  photon, 177

stimulated, 209
emission polarization, 168
emitters, 96, 98, 143–44, 173, 176
   directional, 98
   excited, 173
   individual optical, 176
   nanosized, 176
   quantum dot, 96
end facet, 214–15, 217
energy attenuation, 162
energy conservation, 30
energy dissipation, strong, 138
energy flow, 29–30, 116
energy loss, 162, 192
enhancement factor, 56–57, 60–64, 94
   averaged SERS, 69
   high, 56
   total, 56
etching techniques, 216
evanescent field, 130, 139–40, 159
excitation, 87, 89–92, 112, 115–16, 139–40, 147, 149–50, 158–61, 163–64, 168–69, 172–73, 175–76, 184, 192, 209–10
   counterbeam, 130
   electronic, 141
   experimental, 147
   localized, 86
   longitudinal, 89
   optical, 141
   plasmon, 63, 73, 77
   polarized, 147, 169
   quadrupole, 92
   remote, 81, 174
   resonant, 87, 110
excitation light, 139, 143, 149, 152, 155–57, 159–60, 162, 165–66, 169, 174, 177, 179, 181, 183, 194
excitation polarizations, 166, 180
excitation wavelength, 72, 113, 147, 153–54, 165–66
   free-space, 87
expansion, 23, 25–26
expansion coefficients, 26, 29, 41
exponential decay, 161–62, 220
extinction, 7–8, 30, 49–51, 66–68, 103, 209
   average, 48
   far-field, 67

Fabry–Pérot cavity, 87
Fabry–Pérot resonances, 159
Fano interference, 45
Fano resonances, 16, 67
far-field polarization, 99, 103
far-field radiation, 86
   tunable, 87
FDTD, *see* finite-difference time-domain
FEM, *see* finite element method
Fermi level, 79
Fermi occupation number, 60
Fermi surface, 4
field confinement, 1, 15–16, 192
   electric, 131
   subwavelength, 138
field distribution
   asymmetric, 164
   local, 58, 89
   plasmon, 149, 179
   spatial, 70
   SPP, 157
   symmetric, 157, 181
   zigzag, 182
figure of merit (FoM), 170
finite-difference time-domain (FDTD), 2, 47, 49–51, 58, 111, 176
finite element method (FEM), 2, 47, 49–51

fluorescence, 85, 94, 140, 143–44, 149, 160–61, 173–74, 186, 207
    surface-enhanced, 15, 94, 201
FoM, *see* figure of merit
Förster dipole–dipole interaction, 208
forward scattering, strong, 127
Fourier images, 152–53, 165–66
Fourier imaging, 151–52, 164
Fourier plane, 164–65
Fourier transform, 58
Fourier-transform-based algorithm, 158–59
Fourier transform domain, 158
frequencies, 4, 10, 13, 17, 56–58, 77, 159, 171, 174, 183, 203, 211
    bulk plasmon, 3, 11
    cutoff, 214
    infrared, 94
    operation, 86
    plasmon resonance, 66
    resonant, 91, 206
    spinning, 120
    vibrational, 56
    visible, 13, 93
Fresnel's law, 4
full-width at half-maximum (FWHM), 10, 69, 89–90, 129, 209
FWHM, *see* full-width at half-maximum

GaAs quantum well, 141
gain
    critical, 203
    high material, 214
    saturated internal, 214
gain coefficients, 204, 207, 212, 215, 218–21
    critical, 202, 206
gain compensation, 204
gain layer, 204–5
gain materials, 16, 202, 204–6, 212, 216, 219, 221
gain medium, 202, 205–7, 209, 211–12, 220
gain saturation, 206, 212, 214
gate
    AND, 187
    Boolean logic, 189
    logic, 16, 189–91
    NOR, 190–91
    NOT, 187, 190
    OR, 187, 190
    XOR, 187
Gaussian beam, 32, 114, 116
    paraxial, 147
    polarized, 173
Gaussian laser focus, 125
Gaussian profile, 89–90, 116
GDM, *see* Green dyadic method
generalized Mie theory (GMT), 32, 51, 58–59, 74, 76, 99, 111
glass substrate, 127, 141, 148–49, 151–53, 155, 167
GMT, *see* generalized Mie theory
GNP, *see* gold nanoparticle
gold film, 127–28, 130–31
gold nanobars, 87, 96
gold nanodisks, 128–29
gold nanopads, 127–28
gold nanoparticle (GNP), 8–9, 80, 116, 121–123, 129
gold nanorod, 48–49
gold nanosphere, 64, 172
gold substrate, 68, 70
gratings
    cascading, 185
    cascading corrugation, 185
Green dyadic method (GDM), 46–52
Green function, 2, 47
Green tensor, 47, 49

group velocity, 139, 158–59

half adder, 189
half-wave plate, 65, 102
Hamaker constant, 121
Hankel function, 24, 146
heat-assisted magnetic recording, 15
Helmholtz equations, 27
hole-mask colloidal lithography, 103
hot spots, 45, 57, 73, 79–81, 93, 96, 99, 103, 112, 124–25, 174–75
  quasi-isotropic, 101
hybridization, 65–66, 90, 192
hybrid plasmonic modes, 213–14, 217, 219
hybrid plasmonic waveguide mode, 214
hybrid SP modes, 202–3, 219
  waveguided, 219

incident field, 4–5, 28, 30, 47, 51, 63, 114
incident laser, 61, 114, 116, 150, 170, 191
incident light, 10, 12, 37, 40, 44, 48, 56–57, 71, 74, 77, 147, 172–73, 177, 179, 182
  polarized, 168, 170
incident photons, 115, 120, 172–73
incident polarization, 9, 62, 66, 70, 90, 98–100, 147, 155–56, 177, 179–81, 191, 194
input light, 148, 177, 180, 186–87
intensity, 13, 64–65, 99–100, 103, 126, 128, 131, 143, 146, 154, 159–62, 164, 166–67, 169, 187–88
  emitting, 143
  far-field, 98
  fixed, 159
  near-field, 67, 178, 186
  normalized, 100
  relative, 170
  strong, 89
  threshold, 187, 189, 214
  transmitted, 127
intensity enhancement, 34, 81, 93
  far-field, 103
interaction
  dipole–dipole, 113
  interparticle, 78, 132
  NW–emitter, 139
  plasmon–exciton, 176
  spin–orbit, 17, 171, 182
interband transitions, 4, 8, 51, 89
interference, 29, 67, 116, 168, 186
  constructive, 97, 186–87, 189
  destructive, 187, 189

JC, *see* Johnson and Christy
Johnson and Christy (JC), 4, 7–8, 39, 46, 49, 51, 64

Kretschmann configuration, 12, 139, 143, 158

Laplace equation, 206
laser, 100, 110–11, 116, 118, 124–25, 127–29, 160, 163, 174–75, 180, 208–9, 211, 214–15, 217, 221
laser beams, 110, 114, 116, 121, 124–25, 144, 178, 184–85, 188
laser excitation, 150, 173–74
  pulsed, 90
laser light, 80, 141–45, 147, 150, 153, 164, 167, 173, 175, 178, 183, 190
  focused, 139, 141–43, 147
  polarized, 155

laser polarization, 64, 113, 116–20, 122–23, 150, 166, 168, 178
laser tweezers, 110, 126–27, 132
lasing, 16, 203, 207, 209, 212, 215–18, 220
lasing threshold, 202, 214–15
  low, 215, 217
lattice constant, 22, 48, 78
LCP, *see* left-circularly polarized
leakage radiation, 141–42, 148, 152, 166
LED, *see* light-emitting diode
left-circularly polarized (LCP), 103–4, 120
left-handed chirality, 155
LF-CDA, *see* Lorentz force–coupled dipole approximation
light-emitting diode (LED), 141
light–matter interaction, 96, 194
light polarization, 98
  white, 117–18, 120, 122–23
light scattering, 7, 21–23, 51, 139, 142
  elastic, 22
  inelastic, 22
linear response theory, 59
  first-principle, 58
local electric field, 29, 92, 94, 100, 143, 191
  enhanced, 174
  strong, 129
local field, 29, 45, 57, 89, 91, 95, 101, 141, 143–44, 177, 187
  enhanced, 175
  magnified, 56
  strong, 191
localization, 104, 132, 138, 141, 212
localized surface plasmon (LSP), 57, 62, 77, 80–81
localized surface plasmon resonance (LSPR), 14–16

logic gates, 16, 189–91
long-range surface plasmon polariton (LRSPP), 203–5
Lorentz force, 111, 113–14
Lorentz force–coupled dipole approximation (LF-CDA), 113–14
LRSPP, *see* long-range surface plasmon polariton
LSPR, *see* localized surface plasmon resonance
LSP, *see* localized surface plasmon

Mach–Zehnder interferometer, 193
magnetic field, 24, 26, 28, 58, 146, 171, 211
malachite green isothiocyanate (MGITC), 69–70, 175
manipulation
  noninvasive, 132
  optical, 86
matrix inversion, 40, 44, 68
Maxwell's equations, 2, 10, 22, 27, 49, 58, 171
Maxwell's stress tensor, 111, 113–14, 124
Maxwell's stress tensor and Mie theory (MST-Mie), 113–14
metal–insulator–semiconductor interface, 202–3, 213, 221
metallic nanostructures, 85–86, 93, 96, 105, 207–8
metal nanodisks, 103, 127
metal nanoparticles, 5, 7, 16, 60, 69, 72–73, 110–11, 114, 116–17, 121, 124, 126, 129, 132
  asymmetrical, 121
  colloidal, 56
  finite, 71
  near-field-coupled, 112
  planar, 103

spherical, 5, 110–11, 118
trapped, 110, 117
metal nanostructures, 2, 15–16, 93, 110–11, 127–28, 130, 132, 174
metal nanowaveguides, 192
metal NWs, 71, 138–41, 145, 147, 149, 159, 163, 168, 170–71, 173, 176, 193–94
MGITC, *see* malachite green isothiocyanate
Mie coefficients, 28, 41
Mie scattering, 22
Mie theory, 2, 23, 39, 46–47, 50–51, 58, 80, 113–14, 124
MLWA, *see* modified long-wavelength approximation
mode confinement, 203, 212, 215, 217
modified long-wavelength approximation (MLWA), 7, 118
MST-Mie, *see* Maxwell's stress tensor and Mie theory
multinanoparticles, 117
multiple-multipole method, 46
multipoles, 38–39, 46, 68, 113
electric, 62

NW–nanoparticle junction
nanoaggregates, 86
nanoantennas, 16, 71, 86–87, 93, 95–96, 99–103, 105, 129, 140
bimetallic, 98
gold bowtie, 94
gold dipole, 129
metallic bowtie, 94
plasmonic dimer, 96
triangular, 90
nanocircuits, 182, 192, 194
nanodisks, 77, 86, 111, 128
nanofiber, 159, 192–93
nanogaps, 15, 57, 60–71, 73–82, 98, 111, 174–75, 182

nanoholes, 15, 72–73, 127–28
nanolithography, 15
nanoparticle, 57–60, 62–63, 65–66, 68–69, 71–73, 75–78, 81–82, 99–101, 138–40, 142, 157, 177, 181–82, 206–7, 209–10
aggregated, 60, 76, 81
anisotropic, 120
aspherical, 2
core-shell, 209
dielectric/metal, 110
disk-shaped, 97
faceted, 63
metal-dielectric core-shell, 202
metallic, 87, 110, 130, 206–7
self-assembled, 121
nanoparticle aggregates, 76, 81, 105, 126
nanoparticle dimers, 60, 67, 69–70, 73, 99
nanoparticle–nanohole structure, 72–73
nanoparticle–film structure, 72–73
nanophotonic devices, 17, 194
nanorods, 2, 86, 95, 117–20, 122, 132
nanoshells, 73–74
nanospectroscopy, 194
nanosphere plasmons, 67
nanospheres, 7, 34, 45, 57, 59–60, 63, 78–79, 101–2, 171–72
nanostructures, 2, 14–15, 58–59, 86–87, 96, 103, 105, 110, 119, 168, 171
nanowaveguides, 16
nanowire (NW), 69, 71, 119–20, 138, 141, 143–44, 146–48, 150–53, 155–56, 158–60, 164–67, 173, 175, 177, 189, 192, 213–15
near-field coupling, 12, 68, 104, 110, 117, 122–25, 129–30, 144

near-field distribution, 89–90, 96, 143, 149, 153–54, 177, 180–81, 184, 187–88, 191–92
near-field pattern, 91, 95, 105, 149, 153, 155
near-field scanning optical microscopy, 86
near-infrared (NIR), 96, 117–18, 121, 124–26, 132, 193, 211
Neumann function, 24
NIR, *see* near-infrared
nitrogen vacancy, 140, 176
noble metals, 2, 4, 13–14
NOR gate, 190–91
NOT gate, 187, 190
NW, *see* nanowire
NW–emitter system, 173
NW–nanoparticle junction, 175, 177
NW–nanoparticle structure, 142
NW–nanoparticle system, 175, 178
NW networks, 157, 177–78, 181, 186–89, 194
NW plasmons, 141, 173
NW–QD system, 174
NW SPPs, 141, 162, 174
NW structures, 177–79, 181, 183–85, 188, 190, 194
NW waveguide, 139–41, 160, 162, 175

ohmic heating, 31
ohmic loss, 16, 29, 146, 159, 202–3
oil, index-matching, 183
oil immersion, 141, 152, 155
optical fiber, 138, 140, 143, 160
optical fields, 94, 105, 138, 170
optical forces, 16, 109–14, 117, 124, 130–31
  plasmon-assisted, 110, 117
  plasmon-induced, 112
  surface-plasmon-assisted, 132
  surface-plasmon-induced, 121
optical frequencies, 2, 12, 14, 85, 87, 93, 192
optical gain, 212, 218
optical microscopy, 94, 142, 144, 171
optical near-field coupling, 93
optical trap, 110–11, 114–15, 117, 119, 121, 125
optical trapping, 109–10, 115–16, 119, 124, 127–28, 132
optical tweezers, 109, 117, 124, 126–28
orbital momentum density, 172
OR gate, 187, 190
oscillations, 1, 56, 86–87, 117, 159
oxidation-reduction cycles, 55

parallel incident polarization, 168
parallel polarization, 113, 149, 166, 168–69
particle plasmon resonance, 10
particle plasmons, 10
pentamer, 66–68
periodic zigzag patterns, 172–73, 183
permeability, 22, 27, 51, 58
permittivity, 11–12, 27, 58–59, 162, 171, 203
  complex, 206
  frequency-dependent, 51, 139
  frequency-dependent complex, 2
  local, 59
  local macroscopic, 58
  relative, 10
  wavelength-dependent, 97
perpendicular plasmon mode, 123
perpendicular polarization, 70, 113, 166, 168
phase delay, 119–20, 155–56
  optical, 186
photocurrent, 144

plasmon-induced, 144
photodetectors, 16, 143, 145, 208
photoluminescence, 80, 220
photon cavity mode, 217
photon couplers, 98
photons, 9, 17, 21–22, 32, 56–57, 109, 137, 139, 141–42, 150–52, 164–65, 171–73, 176, 183, 193
 circularly polarized, 172
 emitted, 142, 168
photon scanning tunneling microscope (PSTM), 143
plane wave, 9, 11, 23, 40, 69–70, 73, 114, 119–20
 circularly polarized, 172
 linearly polarized, 9, 25, 45, 121
 monochromatic, 23
plasmon-based integrated photonics, 211
plasmon-based nano-optics, 140
plasmon eigenmodes, 148
plasmon-enhanced spectroscopy, 15
plasmon fields, 143, 149, 177–78
plasmon hybridization, 66–67, 72, 77–79, 122
plasmonic antennas, 85–87, 89, 105
 coupled, 86, 94
plasmonic circuits, 1, 138, 153, 177, 192, 202, 218–19, 221
plasmonic devices, 16, 139, 194, 202, 221
 gain-assisted, 203
plasmonic laser, 213–15, 220
 hybrid, 215, 217–18
 large-scale integrated, 216
 low-threshold hybrid, 215
 nanometer-scale, 213
 red-emitting hybrid, 216
plasmonic nanoantennas, 1, 96
plasmonic nanocircuits, 183
plasmonic nanostructures, 15, 17, 171
plasmonic NW, 144, 157, 173, 192, 194
plasmonics, 1, 14, 17, 71, 130, 138, 147, 193–94, 214
 chiral, 16
 graphene, 17
 quantum, 16
plasmonic signal, 202–3, 218
plasmonic waveguides, 16, 138, 176, 189, 194, 202, 205, 213, 218–20
plasmon interference, 185, 187
plasmon lasers, 214, 216
plasmon modes, 66, 138, 141, 147–49, 151–52, 154, 156–57, 164, 168, 181–82, 209, 214
 discrete, 71
 hybrid, 215
 leaky, 141
 longitudinal, 87
 longitudinal dipole, 87
 lowest-order, 147
 multiple, 189
 quadrupolar, 91–92
 resonant, 89–90, 92
 transverse dipole, 87
plasmon propagation, 130, 138, 149
plasmon resonance, 73, 207
plasmon routing, 178–79, 181, 187–88, 191
plasmons, 4, 57, 67, 71, 73, 140–41, 143, 145, 150, 155, 162–63, 174, 176–78, 186–87, 192–93
polarization, 45–46, 48, 61, 63–70, 76–77, 99–101, 116–22, 149–50, 155–57, 167–70, 177–81, 183–86, 188, 191–93, 219

circular, 120
elliptical, 120
fixed white-light, 118
horizontal, 63
linear, 117–19, 125, 184
longitudinal, 90
macroscopic, 3
mode, 216
optimal, 186
orthogonal, 99, 219
tunable, 219
polarization dependence, 61, 65, 70, 122, 177, 180
polarization-maintaining functions, 168
polarization-rotating functions, 168
polarization rotation, 76–77, 101–2
   wavelength-dependent, 77, 100, 102
Poynting intensity, 165
Poynting vector, 30
prism coupling, 12
propagating plasmons, 149–50, 152, 158–59, 166, 173, 175–76, 193
propagating SPPs, 139, 141–44, 149, 163–64, 168, 173–74, 176, 182–83, 193–94, 213, 218
propagating SPs, 10, 58, 96, 212
propagation, 12–14, 16, 110, 119, 130, 138–39, 146–47, 160–61, 167–68, 192, 194, 203–4, 212
   chiral, 155
   gain-assisted plamonic, 221
   helical, 155–56
   light, 192
   long-distance, 14, 161
   long-range, 1
   lossless, 212
   SP, 16
   transverse, 173
   zigzag, 155
propagation loss, 160, 162–63, 192, 203–6, 213, 215, 218–21
PSTM, *see* photon scanning tunneling microscope
pump-probe technique, 203, 218–19
Purcell effect, 214, 221
Purcell factor, 214–17

QD, *see* quantum dot
QD emission, 140, 174
QD fluorescence, 144, 149–50, 154–55, 157, 167, 178, 180–81, 186–88, 191–92
QD fluorescence imaging, 157, 177, 191
quantum cascade laser, 141, 211
quantum dot (QD), 32, 96, 140, 144, 149, 154–55, 160, 173–74, 176–77, 207, 212
quantum emitters, 85, 140, 173, 176
quantum interference, 17
quantum mechanics, 24–25
quantum nanophotonic circuits, 17
quantum optics, 17, 176
quantum photonics, 194
quarter-wave plate, 183–84
quasi-TM mode, first-order, 131

radiation, 6, 97, 105, 109, 141, 151–53, 159, 164–67, 206, 208, 221
   intense pulsed, 89
   leakage, 141, 148, 163, 167
   leaky, 167
radiative damping, 7, 9, 13, 31
Raman/Brillouin scattering process, 22
Raman emission, 56, 65, 81

Raman enhancement, 69–70
Raman excitation, 124
Raman image, 175
Raman intensity, increased, 55
Raman probes, 124, 126
Raman process, 56
Raman scattering (RS), 15–16, 55–56, 68, 76, 80, 98–103, 111, 143, 174–75, 201
Raman shift, 99, 126
Raman signal, 45, 55–56, 99
Raman spectra, 56, 60, 70, 85, 125
Rayleigh scattering, 22, 129, 207
RCP, see right-circularly polarized 103
red shift, 6–8, 122–24, 129
refractive index, 12, 27–28, 33, 76, 146–49, 153, 155, 162, 164–65, 172, 182, 204, 212, 215
remote excitation, 81, 174–75
remote excitation SERS, 71, 175
remote SERS, 175
resonance, 6–7, 11, 13–14, 32, 40, 57, 67, 87, 92–93, 111, 122, 127, 206–7
 bright plasmon, 45
 dark plasmon, 45
 localized, 211
 localized SP , 203
 longitudinal, 87
 longitudinal plasmon, 87–89
 plasmon, 6
 quadrupolar, 66, 91
 quadrupole, 116
 surface plasmon, 1, 14, 31, 86, 122
right-circularly polarized (RCP), 103–4
room temperature, 111, 204, 215–16
routers, 16, 177, 202
routing, 176, 179, 188
RS, see Raman scattering

scalability, 218
scanning electron microscopy (SEM), 60, 73, 89, 129, 152
scanning near-field optical microscopy (s-SNOM), scattering-type, 94
scanning near-field optical microscopy (SNOM), 90–92, 138, 143, 171
scanning tunneling microscope (STM), 141, 143
scattering, 1–2, 6–9, 22–23, 27, 29–31, 37–41, 43–45, 68, 97–98, 116–17, 140, 172, 180–82, 186, 190–91
 inward, 116
 outward, 116
scattering coefficients, 32–33
scattering force, 110–11, 115–16, 130
scattering intensity, 118, 186, 188, 207
scattering spectroscopy, far-field, 86
scatterers, 23, 27, 30–31, 47–48, 51
Schwinger equation, 47
second harmonic generation (SHG), 96
SEF, see surface-enhanced fluorescence
SEIRA, see surface-enhanced infrared absorption
self-induced back-action (SIBA), 127
SEM, see scanning electron microscopy
semiconductor, 174, 192, 202, 212–13, 215, 221

semiconductor nanowires, 118
semiconductor QDs, 143
semiconductor quantum wells, 213
sensors, 79, 98, 132, 149, 197, 206
SERS, *see* surface-enhanced Raman scattering
SERS, single-molecule, 45, 57, 60, 71, 80–81
SERS signal, 60, 64, 69, 79, 125–26, 175
SHG, *see* second harmonic generation
short-wavelength light, 161
SIBA, *see* self-induced back-action
signal amplification, 203, 218–19
signals, 143, 153, 191–92, 218
   electric, 138
   high background, 175
   hybrid plasmonic, 202–3, 218
   plasmon, 177
   waveguided, 219
   weak short-range SPP, 220
silica nanofibers, 193
silicon nanocrystals, 141
silver nanoparticles, 10, 29, 31, 88, 112, 114, 124–26, 181
silver nanowires, 119, 138, 143, 147–49, 152, 161, 163, 166, 174, 183, 185, 217
single-mode operation, 219
single-molecule level, 60
single-molecule SERS detection, 61
single-particle level, 121, 132
single-particle spectroscopy, 87
single-photon transistor, 176
SNOM, *see* scanning near-field optical microscopy
SOI, *see* spin–orbit interaction
SP, *see* surface plasmon
   long-range, 203
   long-range propagating, 221
   single quantized, 17

SPASER, *see* surface plasmon amplification by stimulated emission of radiation
SP characteristics, prominent, 15
SP coupling, 101, 105
spectral interferometry, 159
spherical nanoparticles, 2, 59, 112, 119–20, 122, 206–7
spin–orbit interaction (SOI), 17, 171–73, 182
SPP, *see* surface plasmon polariton
   chiral, 155–56, 168
   counterpropagating, 130
   long-range, 138
   voltage-controlled, 141
SPP damping, 138, 163
SPP emission, 160, 164
SPP excitation, 12, 160, 164, 172
SPP group velocity, 158–59
SPP loss, 162–63
SPP modes, 139, 153, 157, 167–68, 191, 194
SPP propagation, 138, 147, 157, 160–61, 163, 177, 194
SPP waves, 155, 168
SPP, *see* surface plasmon propagation
SPR, *see* surface plasmon resonance
   material-dependent, 98
SPR frequencies, 15–16, 87, 96
stable trapping, 117, 130
stimulated emission, 202–3, 207–10, 219, 221
surface plasmon amplification by stimulated emission of radiation (SPASER), 208–9, 213, 215, 220–21
STM, *see* scanning tunneling microscope
Stokes scattering, 56
Stokes shifts, 62, 65

subwavelength confinement, 14, 138, 213
subwavelength energy transport, 45
supercontinuum light, 178, 180
super-resolution, 80, 176
surface-enhanced fluorescence (SEF), 15, 94, 201
surface-enhanced infrared absorption (SEIRA), 15
surface-enhanced molecular spectroscopy, 93
surface-enhanced Raman scattering (SERS), 15, 40, 55–56, 59, 65, 75–76, 80–82, 86, 111, 124–26, 174
surface-enhanced spectroscopic sensing, 117
surface-enhanced spectroscopy, 1, 110
surface plasmon (SP), 1–2, 10, 15–16, 56–57, 87, 100–102, 110, 130, 132, 138, 172–73, 176, 201–3, 206–8, 210, 212, 219, 221
surface plasmon polariton (SPP), 10–14, 130, 137–45, 147–48, 155–60, 163, 165, 167–68, 170, 172, 174, 177–78, 183–85, 193–94, 202–4, 212–13, 215–18, 220–21
surface plasmon resonance (SPR), 1, 14–16, 31, 86–87, 96, 98, 122

TDLDA, *see* time-dependent local density approximation
TERS, *see* tip-enhanced Raman scattering
THG, *see* third harmonic generation

third harmonic generation (THG), 96
time-averaged power flow, 155–56
time-dependent local density approximation (TDLDA), 78–79
time domain method, 51

tip-enhanced Raman scattering (TERS), 15, 68–70, 81, 171
TM, *see* transverse magnetic
T-matrix method, 46
TM modes, 215–16
TM waves, 216
TPL, *see* two-photon-induced luminescence
translation coefficients, 31–32, 35–36, 41–42
transverse magnetic (TM), 24, 33, 146, 215–16
trapping, 110–11, 115–16, 127–29
trapping laser, 110, 117–19, 122, 125, 129
trimer, 44, 76, 86, 99–102, 112
twisted orbital momentum density, 173
two-beam interference, 188
two-photon-induced luminescence (TPL), 89–90, 93, 96, 105

ultrasensitive SERS analysis, 65
ultrashort SPP pulses, 158
ultraviolet (UV), 3–4
UV, *see* ultraviolet

vacuum, 11, 27, 34, 47, 50, 64, 68, 158–59, 171, 203–4
vacuum permittivity, 3
van der Waals attraction, 121, 124
van der Waals energies, 115
van der Waals potential, 114–15, 121

vector spherical harmonics (VSH), 23–28, 31–33, 35, 41, 58–59, 114
viscous hindrance, 119
visibility, 185–86
VSH, *see* vector spherical harmonics

wave front, 157, 182
waveguides, 138, 141, 143, 212, 216, 220
  gold strip, 145
  hybrid, 194
  LRSPP, 204–5
  metal-insulator-metal, 145
  nanostrip, 141
  one-dimensional optical, 138
  photonic, 192, 194
  polymer, 194
  slot, 141
waveguiding, 14, 17
wavelength, 21–22, 61–64, 98–101, 104, 116, 118, 153, 159–62, 165–67, 169–70, 172, 177–80, 184, 192–93, 204–6
  free-space, 13
  half, 71–72
  incident, 34, 39, 46, 62
  long, 178
  multiple, 179, 183
  optical, 94
  resonant, 112
  short, 7, 66, 178
  trapping, 116, 126

ultraviolet, 215
vacuum, 146, 149, 170, 206
wavelength dependence, 99, 178–79
wavelength-dependent directional scattering, 98
wave-particle duality, 17
whispering gallery modes, 218
white light, 118–19, 122, 157–58, 178, 181
white-light polarization, 117–18, 120, 122–23
white-light supercontinuum generation, 16, 93
wide-field excitation, 149–50, 178
wide-field illumination, 192
wire plasmons, 71–72
  long-wavelength, 71
  symmetric, 72

XOR gate, 187
X-ray beam, 21

Yagi–Uda antenna, 96–97
Yagi–Uda gold nanoantenna, 97
Yagi–Uda plasmonic antenna, 96
Yagi–Uda structures, 105
Y-branch structure, 183–84
Yee algorithm, 58
Y-shaped branched NW structure, 183
Y-shaped microfluidic channel, 126

ZnO NWs, 192–93

PGSTL 11/01/2017